LIGAND–RECEPTOR ENERGETICS

LIGAND–RECEPTOR ENERGETICS

A Guide for the Perplexed

IRVING M. KLOTZ
Northwestern University

A Wiley-Interscience Publication
JOHN WILEY & SONS, INC.
New York / Chichester / Weinheim / Brisbane / Singapore / Toronto

This text is printed on acid-free paper.

Library of Congress Cataloging in Publication Data:

Klotz, Irving M. (Irving Myron), 1916–
Ligand-receptor energetics : a guide for the perplexed / Irving M. Klotz.
p. cm.
"February 1997."
Includes index.
ISBN 0-471-17626-5 (cloth : alk. paper)
1. Ligand binding (Biochemistry)—Thermodynamics. I. Title.
QP517.L54K57 1997
574.19′283—dc20 96-34518
CIP

Printed in the United States of America

10 9 8 7 6 5 4 3 2 1

Dedicated to
E. Q. Adams
H. S. Simms
J. A. V. Butler
and others
whose contributions
are no longer
remembered.

CONTENTS

PREFACE

To understand the dependence of a biological response on an effector stimulus, one must acquire a clear understanding of the energetic and molecular principles governing interactions between small molecules and biomacromolecules. The first step in essentially all biological activities is an interaction between separate molecular constituents, ligand and receptor, to form a macromolecular complex. Such interactions play a vital role throughout the basic life sciences, in biochemistry, biophysics, pharmacology, physiology, immunology, endocrinology, neurobiology, molecular biology, and cell biology.

New investigators, as well as experienced ones, are often unaware of the origins of the fundamental concepts and theoretical procedures in ligand–receptor energetics so they fail to recognize hidden assumptions and premises.

The objective of this volume is to present the core principles that provide the foundation for quantitative perspectives. These provide a framework for interpreting experimental observations in a wide range of ligand–receptor interactions that pervade the basic life sciences.

LIGAND–RECEPTOR ENERGETICS

INTRODUCTION

The first step in essentially* all biological activities is a union between separate constituents into a merged entity. Such combinations play a cardinal role throughout the life sciences. This fundamental feature of biological phenomena is expressed in the maxim (attributed to Paul Ehrlich) *Corpora non agunt nisi ligata* (a substance is not effective unless it is linked to another).

At the molecular level, if one of the participants in a united entity is smaller than its complementary partner, the former is usually denoted as the *ligand* and the latter as the *receptor.* Thus in an enzyme–substrate complex, the substrate is the ligand, and the enzyme is the receptor. In blood, oxygen, metabolites, and hormones are the ligands; carrier proteins are the receptors. In immunological interactions, the ligand is the hapten or antigen; the receptor is the immunoglobulin. Neurotransmitters or messengers are effector ligands when they are bound to receptor sites at synapses or cell membranes.

Representative specific ligand–receptor complexes encountered in the basic life sciences include the following:

*When radiation is involved, the first step is the absorption of a quantum of electromagnetic energy. In an electrical stimulus, the initiating impulse leads to a change in the electromagnetic field.

Receptor	Ligand
Enzyme	*Substrate, Modulator*
Lysozyme	Oligosaccharide
Aspartyl transcarbamoylase	Carbamyl phosphate, cytidine triphosphate
Transporter	*Transportee*
Hemoglobin	Oxygen
Serum albumin	Metabolites, hormones, drugs
Concanavalin A	Trimannoside
Maltoporin	Maltotriose
Antibody	*Antigen*
Anti-DNP antibody	Dinitrophenyllysine
T-cell receptor	Peptide-major histocompatibility complex
Membrane	*Agonist*
Cell	Hormone messengers
Insulin receptor	Insulin
Synapse	*Neurotransmitters*
Electroplax	Acetylcholine
Brain	Morphine, endorphins

If one aspires to carry out a thorough study of biological responses to molecular stimuli, one needs to elucidate the following aspects of ligand–receptor interactions:

1. The number of ligand molecules bound to the receptor under various environmental circumstances.
2. The affinity of the interaction, the strength of the binding of the ligand to the receptor.
3. The forces involved in the interactions.
4. The molecular nature of the interactions and subsequent cellular events.

These features will be scrutinized in the order listed.

CHAPTER 1

NUMBER OF LIGANDED MOLECULES

Binding measurements provide the essential data from which all aspects of ligand–receptor interactions are then examined.

1.1. EXPERIMENTAL DETERMINATION

A vast array of physicochemical and biological methods have been developed to measure and study ligand binding (Fig. 1.1). Each of these techniques depends on one of the following principles:

1. Determination of the concentration of free, unbound ligand.
2. Detection of a change in a physicochemical or biological property of the bound ligand.
3. Detection of a change in a physicochemical or biological property of the receptor or ligand–receptor complex.

1.1.1. Determination of Concentration of Free Ligand

For soluble systems, the simplest and most broadly applicable procedure is classical equilibrium dialysis (1,2). The basic premise of this method is illustrated schematically in Figure 1.2. A solution of protein or other biomacromolecule on one side of a semipermeable

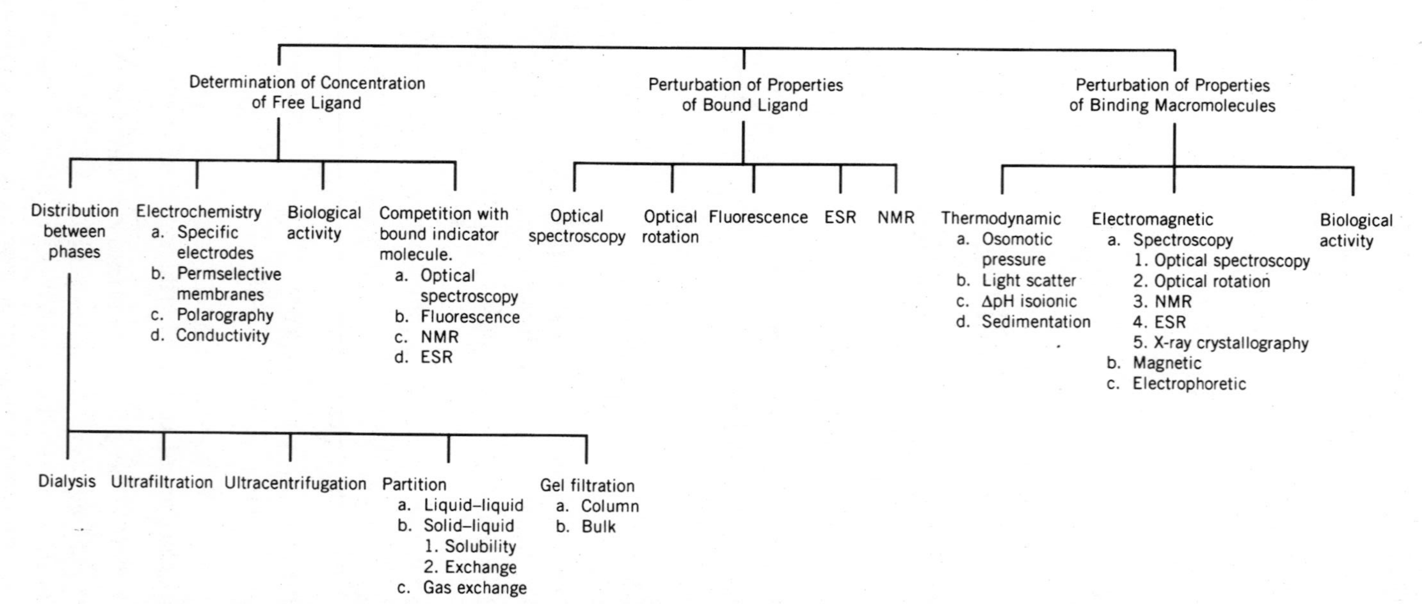

Figure 1.1 Methods of studying the binding of small ligands by macromolecules.

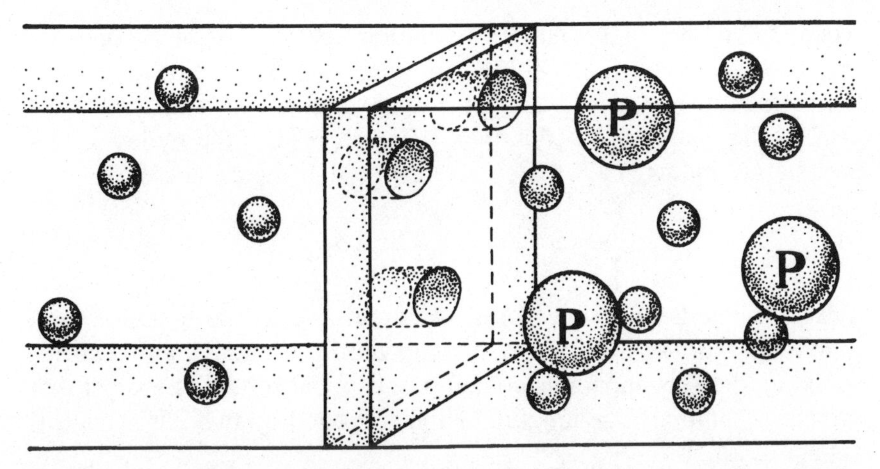

Figure 1.2 Schematic illustration of the distribution of a small molecule ligand between two compartments that are separated by a dialysis membrane that is permeable to the ligand (small *spheres*) but not to a protein P or large receptor (large *spheres*).

membrane is allowed to reach equilibrium with a ligand in the same solvent on the other side. At equilibrium, the chemical potential, and hence the thermodynamic activity, of the free ligand must be the same in both compartments. The receptor concentration generally is low so that its effect on the activity coefficient of the ligand is negligible. Consequently, the concentration of free ligand should be the same on both sides of the membrane. In principle, then, if one determines the total amount of ligand in the entire container and combines that with a measurement of the free ligand concentration in the receptor-free compartment, one can compute the amount of bound ligand.

The extent of binding B is normally expressed as

$$B = \frac{L_R}{R_T} \tag{1.1}$$

where L_R is the number of moles of ligand on the receptor and R_T is the total moles of receptor. If the total moles of ligand in the container L_T is known and the concentration of free ligand (L) has been determined, then

$$L_R = L_T - V_T(L) \tag{1.2}$$

where V_T is the total volume of solution, the sum of both compartments. In practice, most ligands are also bound to some extent by the membrane of the dialysis apparatus (see Fig. 1.2) and sometimes also by the walls of the container. If L_M represents the moles of this peripheral binding, then the correct expression for L_R becomes

$$L_R = L_T - V_T(L) - L_M \tag{1.3}$$

Two methods have been used to obtain L_M. In the first, for each individual binding experiment, a companion is arranged with a corresponding dialysis apparatus containing identical solutions, except that protein is omitted. The amount of ligand bound by this blank control apparatus L'_M is

$$L'_M = L_T - V_T(L') \tag{1.4}$$

where (L') is the concentration of free ligand in this control. Using equation 1.4 to replace L_T in equation 1.3, we find that

$$L_R = V_T[(L') - (L)] + (L'_M - L_M) \tag{1.5}$$

If the amount of ligand bound by the membrane in the blank L'_M were equal to that picked up when protein is present L_M then

$$L_R = V_T[(L') - (L)] \tag{1.6}$$

This is probably an adequate correction when (L') is near (L), i.e., when the extent of binding by receptor is not large. Nevertheless, it is evident that if binding by the membrane and apparatus were reversible, as it would be for a simple adsorption phenomenon, the uptake by the apparatus with the blank would be larger than that with the protein; for when (L') is greater than (L), as it must be when the receptor binds ligand, then the extent of nonspecific adsorption L'_M will be higher than L_M.

In the second method for making the nonspecific adsorption correction, one arranges a *series* of blanks covering a range of the free ligand concentration (L') that spans the equilibrium concentrations (L) in the presence of receptor. With this series of L'_M values calculated from the battery of controls, one can draw a graph (or prepare a table) showing the variation of nonspecific adsorption with free ligand concentration. Then the appropriate value for L_M to be used in equation 1.3 is that read from the calibration curve for the experimentally determined value of (L), the free ligand concentration in equilibrium with the receptor.

It is also possible to circumvent a correction for nonspecific binding by measuring the *total* ligand concentration in solution in the receptor compartment (see Fig. 1.2) and in the solution in the receptor-free compartment. Then L_R is given by the equation

$$L''_R = V_R[(L'') - (L)] \tag{1.7}$$

where (L'') is the total concentration of ligand, bound and free, in the receptor compartment of volume V_R. Since (L), the concentration of free, nonbound ligand, is the same in both compartments, its value can be obtained from a measurement in the receptor-free compartment. However, there are some drawbacks to this procedure. In general, the attainment of osmotic equilibrium between the two compartments, only one of which has receptor, involves the transfer of some solvent. Hence V_R is not the initial volume of the receptor-containing solution. If one also measures the concentration of receptor (R), at equilibrium, then

$$(R)V_R = R_T \tag{1.8}$$

so, knowing the R_T initially added, one can calculate V_R. In practice, analyzing for concentration of receptor (R) in the presence of ligand, and in fact for (L'') in the presence of receptor, may be experimentally difficult.

A detailed description of the simplest experimental procedure for binding measurements using equilibrium dialysis is given in Appendix A1.

In equilibrium dialysis and in its closely related technique, ultrafiltration, the phases are separated by a semipermeable membrane. Alternatively, the two phases can be mutually immiscible liquids, with the interfacial surface providing the barrier. In gel filtration, the portals to the microchannels maintain the gel interior separate from the bulk solution.

1.1.2. Perturbation of Properties of Bound Ligand

On the receptor a bound ligand experiences a different molecular environment than it does in bulk solution. Thus electronic and other spectroscopic energy levels are perturbed, and spectra of all types are often modified in frequency or amplitude or both. Shifts in optical absorption spectra were used early to measure ligand binding (3). Analogous changes may appear anywhere from the ultraviolet to the infrared to the nuclear magnetic resonance region, as well as in optical rotatory dispersion and fluorescence behavior.

The basic principle for calculating the extent of binding of ligand from spectroscopic perturbations can be illustrated readily with changes in optical absorptions.

Figure 1.3 shows the visible absorption spectra of a sulfonamide dye in a simple buffered aqueous solution and in a corresponding solution to which the protein serum albumin has been added. Clearly, the absorbance of this dye is lowered and shifted in wavelength when protein is present, presumably a manifestation of the formation of a complex. In this particular example, companion equilibrium dialysis experiments established that the sulfonamide is indeed bound by serum albumin.

Under favorable circumstances, the bound ligand will have the

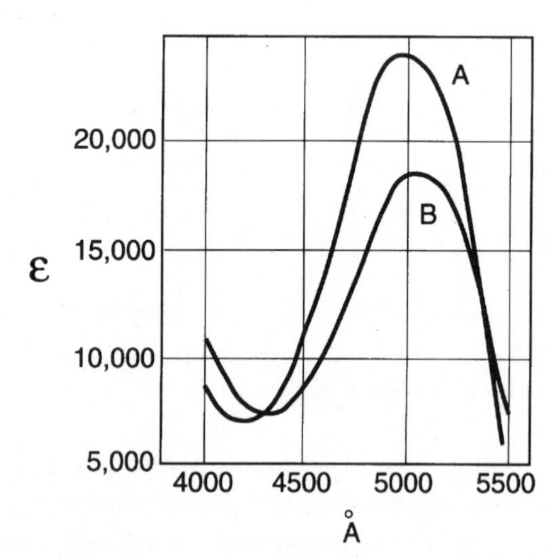

Figure 1.3 Absorption spectra of azosulfathiazole. **A**, in aqueous buffer, at pH 6.9. **B**, In aqueous buffer containing 0.2% bovine serum albumin, at pH 6.9.

same spectroscopic properties no matter how many ligand molecules are linked to the receptor. The absorbance A of light by a solution containing both bound and unbound dye may then be expressed by the relation

$$A = \log I_o/I = \epsilon_f (L_f)d + \epsilon_b(L_b)d \tag{1.9}$$

where I_o is the intensity of the light exiting the solvent; I, the intensity exiting the solution; ϵ, the molar extinction coefficient of the ligand; and d, the optical path length in the absorption cell. The subscripts f and b denote free and bound ligand, respectively. If α_f is the fraction of ligand that is free and (L_T) is the concentration of total ligand, free and bound, then

$$A = \epsilon_f \alpha_f (L_T)d + \epsilon_b(1 - \alpha_f)(L_T)d \tag{1.10}$$

and

$$\frac{A}{(L_T)d} = \alpha_f \epsilon_f - \alpha_f \epsilon_b + \epsilon_b \tag{1.11}$$

If we define

$$\frac{A}{(L_T)d} = \epsilon_{\text{apparent}}$$

then

$$\alpha_f = \frac{\epsilon_{\text{apparent}} - \epsilon_b}{\epsilon_f - \epsilon_b} \tag{1.12}$$

To determine ϵ_f, one need only make absorbance measurements for the ligand in a solution containing no receptor. To determine ϵ_b, one must make absorbance measurements for the ligand in solutions containing increasing concentrations of receptor and find the limiting value as the receptor concentration tends to infinity. (For the sulfonamide dye whose spectra are shown in Figure 1.3, $\epsilon_f = 23,650$ and $\epsilon_b = 17,300$ at 490 nm (4900 Å).) Then one can compute α_f from equation 1.12. Knowing the total ligand concentration (L_T) from the experimental protocol, one can calculate (L_f) and (L_b), and thereafter B, the number of moles of bound ligand divided by the number of moles of receptor, defined in equation 1.1.

When successive bound ligand molecules are in different environments on the receptor, ϵ_b may not be a constant but could be a composite of different molar extinction coefficients for each of the bound ligand molecules in the complex. The number of parameters in equation 1.9 will then need to be increased. Methods for treating such situations have been formulated (4).

If the ligand is a fluorescent molecule and its binding to receptor changes the fluorescence intensity (Fig. 1.4), then one can show that

$$\alpha_b = \frac{I_{\text{apparent}} - I_f}{I_b - I_f} \tag{1.13}$$

where I represents the fluorescence intensity in the solutions denoted by the subscripts. In ligand–receptor fluorescence studies, the intensity

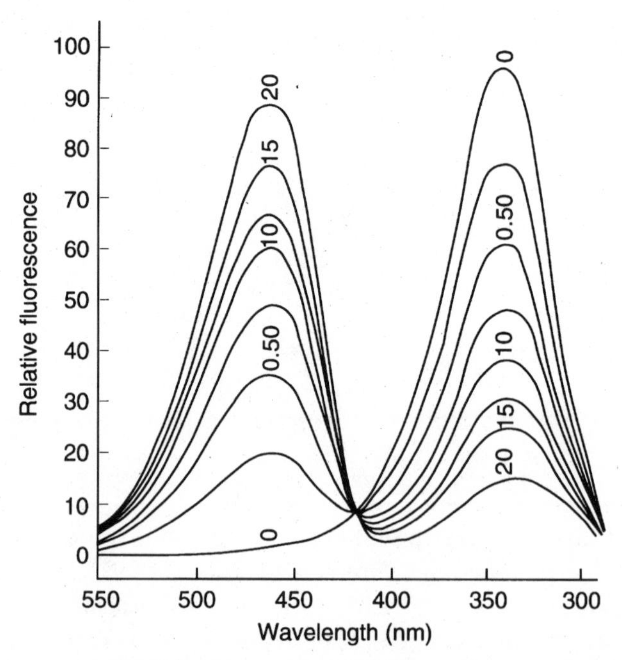

Figure 1.4 Fluorescence intensity of a complex of anilinonaphthalene sulfonic acid with bovine serum albumin at various moles of bound ligand, B. Aqueous buffer, pH 7, 0.1% protein. Data from Ref. 5.

is generally much higher for the bound species, and therefore equation 1.13 is expressed in terms of α_b.

1.1.3. Perturbation of Properties of Receptor

A whole gamut of physicochemical and biological properties of the receptor may be modified by combination with a ligand (see Fig. 1.1). However, these techniques are not generally applicable to quantitative measurements of the extent of binding of ligand because they are sensitive to local structural features of the receptor macromolecule. Nevertheless, they often provide special insights into ligand–receptor complexes. In particular, ligands binding to different sites on a protein will often manifest different responses, so the different sites can be distinguished. This can be a great advantage over equilibrium dialysis, which does not distinguish between different binding sites on a receptor. Methods for detecting the binding of ligands to specific sites on proteins have been described (6,7).

1.2. CONVENTIONAL NOTATION

Over the many decades of experimental measurements of binding of ligands by receptors, different symbols have been used to denote the molecular entities and the measured quantities. We have already made a choice among these in formulating equation 1.1. Since other symbols also appear in the literature to which we shall refer, it will be useful to have a summary of frequent notations.

$$\text{Ligand} \equiv L = l = A \tag{1.14}$$

$$\text{Receptor} \equiv R = P \tag{1.15}$$

$$\frac{\text{Moles bound ligand}}{\text{Moles total receptor}} \equiv B = r = \nu = \frac{L_R}{R_T} \tag{1.16}$$

$$\text{Molar concentration of free (unbound) ligand} \equiv (L) = c = (F) \tag{1.17}$$

Now we can proceed to formulate the theoretical framework in which to express ligand–receptor affinities and then to embark on a conceptual analysis of published data.

REFERENCES

1. W. A. Osborne, *J. Physiol.*, **34**, 84 (1906).
2. I. M. Klotz, F. M. Walker, and R. B. Pivan, *J. Am. Chem. Soc.*, **68**, 1486 (1946).
3. I. M. Klotz, *J. Am. Chem. Soc.*, **68**, 2299 (1946).
4. A. Knudsen, A. O. Pedersen, and R. Brodersen, *Arch. Biochem. Biophys.*, **244**, 273 (1986).
5. E. Daniel and G. Weber, *Biochemistry*, **5**, 1894 (1966).
6. T. E. Creighton, *Protein Function: A Practical Approach*, IRL Press, Oxford, UK, 1989.
7. D. J. Winzor and W. H. Sawyer, *Quantitative Characterization of Ligand Binding*, John Wiley & Sons, Inc., New York, 1995.

CHAPTER 2

AFFINITIES: FROM A SITE PERSPECTIVE

Affinities are assessed quantitatively by the change in free energy ΔG for the union of the constituent entities. In principle, there are several experimental approaches to the evaluation of ΔG (1). In practice, for ligand–receptor complexes only one is accessible: the determination of equilibrium constants.

When several or many ligands can be bound by a receptor, multiple equilibria are established. These can be described by different types of equilibrium constants, which reflect different perspectives in visualizing the equilibria. We shall start by focusing on the sites of the receptor to which the ligands become attached. Our approach will build up from simple to more complex situations.

2.1. SINGLE BINDING SITE: UNIVALENT RECEPTOR

A univalent combination can be written as

$$R + L = RL \tag{2.1}$$

where R represents the receptor and L the ligand (Fig. 2.1). For these species in equilibrium, we can define a *site* equilibrium constant k, by the equation

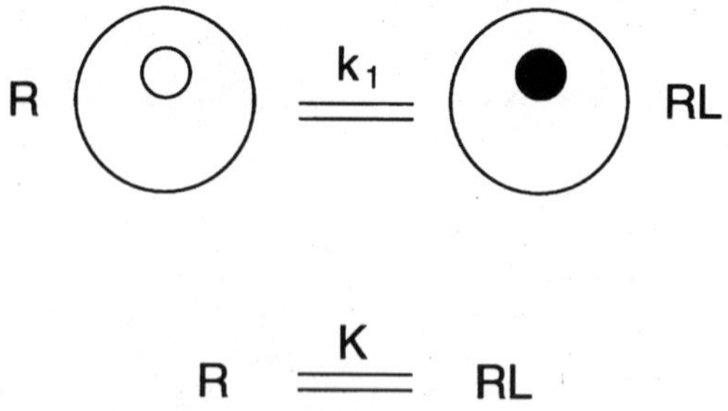

$$R \overset{k_1}{\rightleftharpoons} RL$$

$$R \overset{K}{\rightleftharpoons} RL$$

Figure 2.1 A receptor with a single site for binding of ligand.

$$k = \frac{(RL)}{(R)(L)} \tag{2.2}$$

from which it follows that

$$(RL) = k(R)(L) \tag{2.3}$$

The parentheses represent the concentrations of the respective species.

For an equilibrium in a reaction such as that in equation 2.1 involving 1 : 1 stoichiometry, one can also assign a *stoichiometric equilibrium constant* K

$$K = \frac{(RL)}{(R)(L)} \tag{2.4}$$

Clearly for a univalent combination, where the receptor has a single binding site

$$K = k \tag{2.5}$$

As soon as we treat two-site (Fig. 2.2) or multisite receptors, however, equation (2.5) is no longer valid.

Now we can form an equation to express the moles of bound ligand B as a function of free ligand concentration:

$$B = \frac{\text{moles bound ligand}}{\text{moles total receptor}} = \frac{(RL)}{(R) + (RL)} \tag{2.6}$$

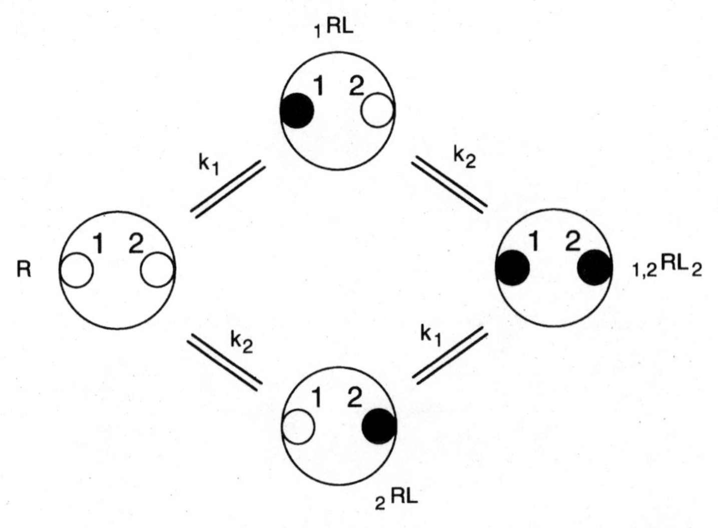

Figure 2.2 A two-site receptor, each site with a different fixed, unchanging affinity, unaffected by the extent of binding of ligand at the other site on the receptor.

Replacing (RL) by the right-hand side of equation (2.3) we find, after canceling the factor (R) that appears in numerator and denominator, that

$$B = \frac{k(L)}{1 + k(L)} = \frac{K(L)}{1 + K(L)} \tag{2.7}$$

Actual experimental data for B and (L) in a monovalent combination can be fit to this equation by a straightforward computer program, and then K can be evaluated. For example:

Myoglobin + O_2 = Mb · O_2, $K \simeq 0.5\ \text{mm}^{-1}$
Chymotrypsin + peptide = ChT · peptide, $K \simeq 10^4\ \text{M}^{-1}$

In the literature a number of different personal names have been appended to equation 2.7. Actually, it was first derived at least as far back as 1913, by Michaelis and Menten (2), the latter being one of the first women biochemists.

It is often convenient to present experimental binding data in graphical form. The most obvious format is a plot of B versus free ligand concentration (L) (Fig. 2.3). An appropriate value of K can be

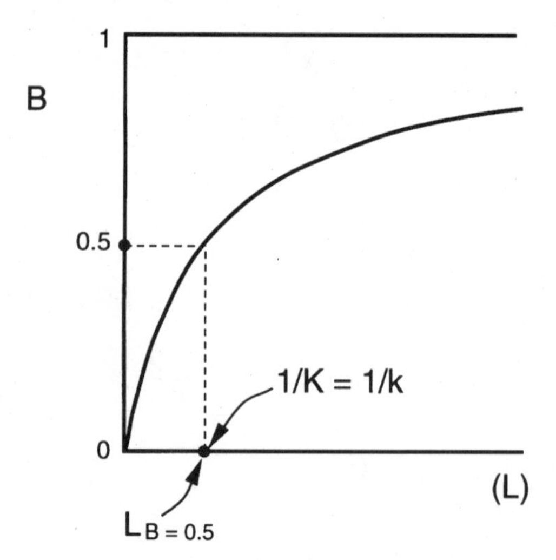

Figure 2.3 Direct plot of moles of bound ligand as a function of free ligand concentration (L) for a single binding site.

obtained from a visual estimate of $L_{B=0.5}$, the value of (L) at which $B = 0.5$. One can confirm that this procedure is valid by inserting into equation (2.7) the values $B = 0.5$ and $(L) = L_{B=0.5}$ and solving for K.

2.2. MULTIVALENT UNIFORM RECEPTOR: MANY IDENTICAL BINDING SITES WITH INVARIANT IDENTICAL AFFINITIES

2.2.1. Algebraic Representation of Binding

Since each site has a fixed, unchanging affinity, each is unaffected by the extent of binding of ligands at other sites on the receptor (Fig. 2.4). Hence for the individual sites we may write

$$B_1 = \frac{k_1(L)}{1 + k_1(L)} \tag{2.8}$$

$$B_2 = \frac{k_2(L)}{1 + k_2(L)} \tag{2.9}$$

$$\vdots$$

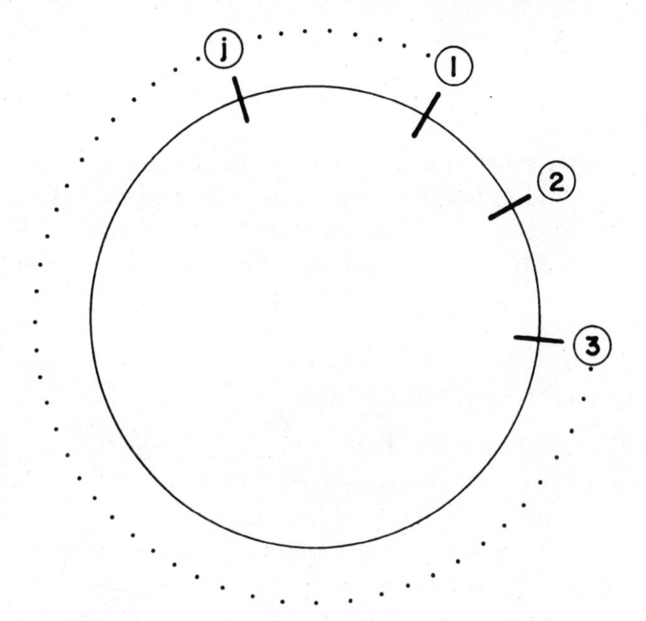

Figure 2.4 Schematic representation of individual sites on a multivalent receptor.

$$B_j = \frac{k_j(L)}{1 + k_j(L)} \tag{2.10}$$

$$\vdots$$

$$B_n = \frac{k_n(L)}{1 + k_n(L)} \tag{2.11}$$

where n is the total number of binding sites on the receptor. For the receptor as a whole,

$$B = B_1 + B_2 + \cdots + B_j + \cdots + B_n \tag{2.12}$$

Since each site is identical with the others and invariant in affinity,

$$k_1 = k_2 = \cdots = k_j = \cdots = k_n \equiv k \tag{2.13}$$

and

$$B = n\,\frac{k(L)}{1 + k(L)} = n\,\frac{K(L)}{1 + K(L)} \tag{2.14}$$

Thus B equals n times the binding of ligand by a single isolated site, for which $k = K$. Here, too, the constants k and n should be evaluated by a best fit procedure using data for the binding variables B and (L).

We shall call a multivalent receptor with identical invariant binding sites an *ideal system*.

2.2.2. Graphical Representations

A direct plot of B as a function of (L) is shown in Figure 2.5. The curve obtained is essentially identical to that in Figure 2.3, except that the plateau approaches n, the total number of binding sites, instead of unity. In both figures, the halfway point $B_{n/2}$ can be used to approximate the binding constant k.

Another graphical form, much to be preferred in practice, as will be shown later, plots B against log (L) (Fig. 2.6). For an ideal system, this curve is symmetrical about an inflection point at $B = n/2$. Furthermore, the value of log (L) at $B = n/2$ is equal to $-\log k$ (or $-\log K$), and hence provides a value for the ligand binding constant.

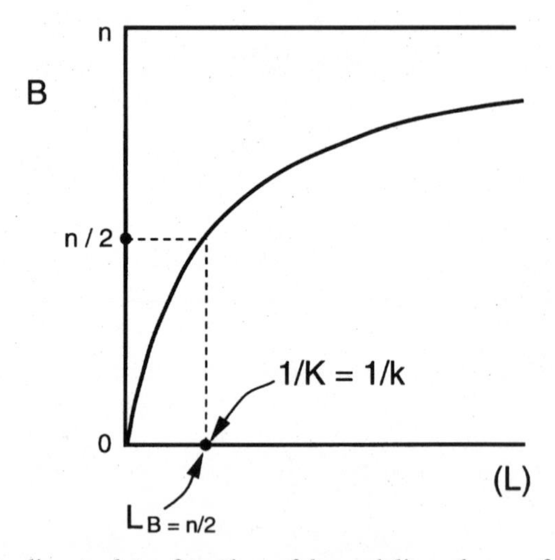

Figure 2.5 A direct plot of moles of bound ligand as a function of free ligand concentration (L) for a receptor with n identical invariant binding sites.

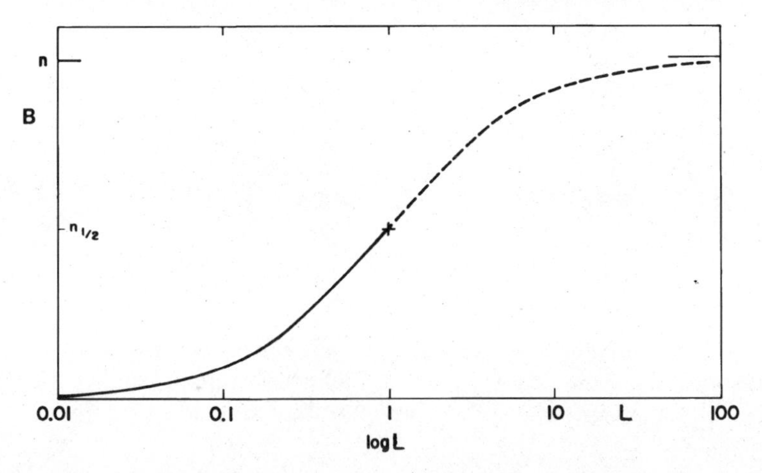

Figure 2.6 Semilogarithmic plot of B as a function of log (L) for an ideal system of n identical invariant binding sites (or of a single site). At the inflection point (+), $B = n/2$ and $\log(L) = -\log k$. The numerical values shown for log (L) on the abscissa are actually normalized ones, i.e., really $\log[(L)/(L)_{B=n/2}]$.

If one moves one log unit to the left of $\log(L)_{B=n/2}$, the value of B is approximately 0.1 n; one log unit to the right, B is approximately 0.9 n. Thus, over a hundredfold range in concentration of free unbound ligand (L), B changes from 10% saturation to 90% saturation. These characteristics serve as a convenient simple criterion for assessing how close a particular real set of data approaches ideal behavior.

Before computers achieved powerful capacities and became universally available, the fitting of actual experimental data to a curve was tedious and time-consuming, even for a relatively simple equation such as (2.14). For almost a century, therefore, intensive efforts were made to find linear transforms for algebraic expressions for curves. For equation (2.14), three such transforms have been developed.

For ideal binding, where equation (2.14) represents the data, each of the linear transforms can be readily derived. If one takes the reciprocal of each side of equation (2.14), one obtains

$$\frac{1}{B} = \frac{1}{nk(L)} + \frac{1}{n} = \frac{1}{n} + \frac{1}{nk}\,\frac{1}{(L)} \tag{2.15}$$

This algebraic form corresponds to a straight line, as is evident in Figure 2.7A.

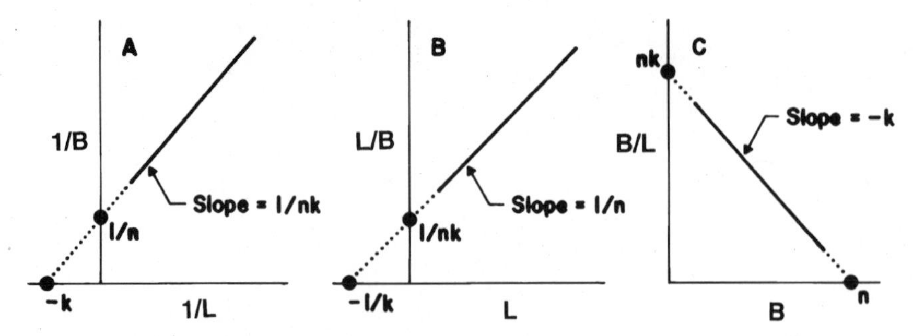

Figure 2.7 Graphical representations of three linear transformations of equation 2.14 for an ideal system of n identical invariant binding sites. The slopes and intercepts of the three straight lines are noted.

A second linear form can be obtained from equation 2.15 by multiplying each side by (L). With minor rearrangement one obtains

$$\frac{(L)}{B} = \frac{1}{nk} + \frac{1}{n}(L) \tag{2.16}$$

the graphical form of which is shown in Figure 2.7B.

A third linear form can be derived by rearranging equation 2.14 to read

$$B[1 + k(L)] = nk(L) \tag{2.17}$$

from which one can obtain

$$B = nk(L) - Bk(L) \tag{2.18}$$

$$\frac{B}{(L)} = nk - kB \tag{2.19}$$

The graphical form is illustrated in Figure 2.7C.

All of these linear algebraic expressions were derived by B. Wolff in the 1920s (3). Specific individual forms were independently discovered or rediscovered by many individuals for application to widely different types of equilibria. As happens commonly in science, proper names have been attached to these equations and to their graphical equivalents, for ease in referring to particular forms. For the graphs in Figure 2.7, the following designations are found in the literature:

A. Double-reciprocal; Langmuir; Lineweaver-Burk.

B. Hanes; Hildebrand-Benesi; Scott.

C. Eadie; Eadie-Hofstee; Scatchard.

In ligand binding studies, the most commonly used name is *Scatchard graph.*

It should be emphasized again that all of these linear forms arise from equation 2.14 and hence are valid only for *ideal* ligand–receptor binding, i.e., when all the sites are identical and invariant in affinity for ligand.

2.3. MULTIVALENT RECEPTOR: SITES WITH DIFFERENT BUT INVARIANT AFFINITIES

Since each site has a fixed, invariant affinity, its binding of ligand is independent of events at other sites (see Fig 2.4). Hence we can return to equations 2.8 to 2.11.

$$B_1 = \frac{k_1(L)}{1 + k_1(L)} \tag{2.8}$$

$$B_2 = \frac{k_2(L)}{1 + k_2(L)} \tag{2.9}$$

$$\vdots$$

$$B_j = \frac{k_j(L)}{1 + k_j(L)} \tag{2.10}$$

$$\vdots$$

$$B_n = \frac{k_n(L)}{1 + k_n(L)} \tag{2.11}$$

For the receptor as a whole,

$$B = \frac{k_1(L)}{1 + k_1(L)} + \frac{k_2(L)}{1 + k_2(L)} + \cdots + \frac{k_j(L)}{1 + k_j(L)} + \cdots + \frac{k_n(L)}{1 + k_n(L)}$$

$$= \sum_{1}^{n} \frac{k_j(L)}{1 + k_j(L)} \qquad (2.20)$$

in which each of the site constants is different.

A special case that is often assumed to be applicable is a system with *two* classes of sites, each with identical invariant affinities that differ from the identical invariant affinities of the second class. Under these circumstances, equation 2.20 can be reduced to

$$B = \frac{n_1 k_1(L)}{1 + k_1(L)} + \frac{n_2 k_2(L)}{1 + k_2(L)} \qquad (2.21)$$

where n_1 and n_2 are the number of sites in the respective classes, and k_1 and k_2 are the respective site binding constants.

When the algebraic expression of either equation 2.20 or 2.21 fits the ligand–receptor binding data, none of the Wolff graphs (see Fig. 2.7) will be linear. For the coordinate pair $B/(L)$ versus B, the Scatchard graph, concave curvature appears (Fig. 2.8).

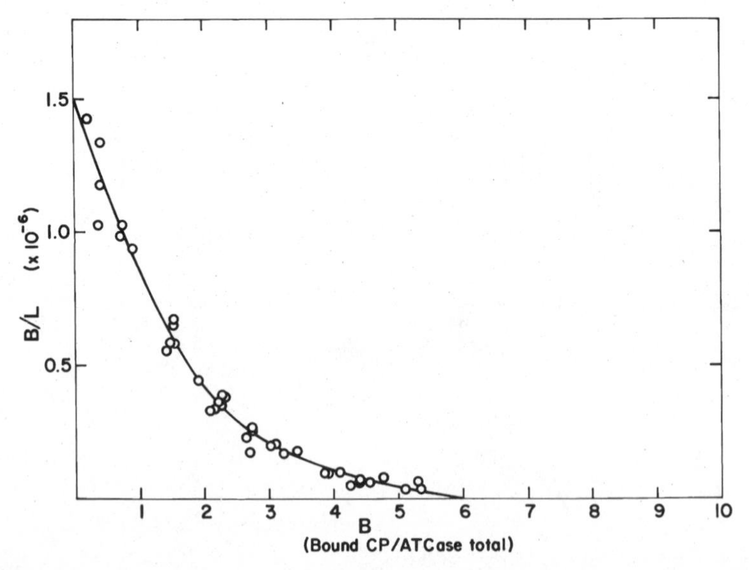

Figure 2.8 Scatchard graph of binding of carbamyl phosphate (CP) by aspartate transcarbamoylase (ATCase). Data from Ref. 4.

As will be demonstrated later, however, the fact that a Scatchard graph can be fitted by the two-site binding equation 2.21 does not mean that the receptor actually has two classes of identical invariant sites. In fact, aspartate transcarbamoylase (ATCase), e.g., the receptor of Figure 2.8, has six identical sites when no ligand is bound. Changes in affinity with increasing occupancy of sites by ligands also lead to curvature in the Wolff graphs, as will be elaborated on later.

2.4. MULTIVALENT RECEPTOR: SITE AFFINITIES CHANGE WITH EXTENT OF OCCUPANCY BY LIGANDS

2.4.1. Divalent Systems

The essence of the complexity that arises when affinities change with occupancy can be illustrated with the simplest multivalent receptor, one with two ligand-binding sites (Fig. 2.9). In this situation there are four different site binding constants. (It will be shown later that only

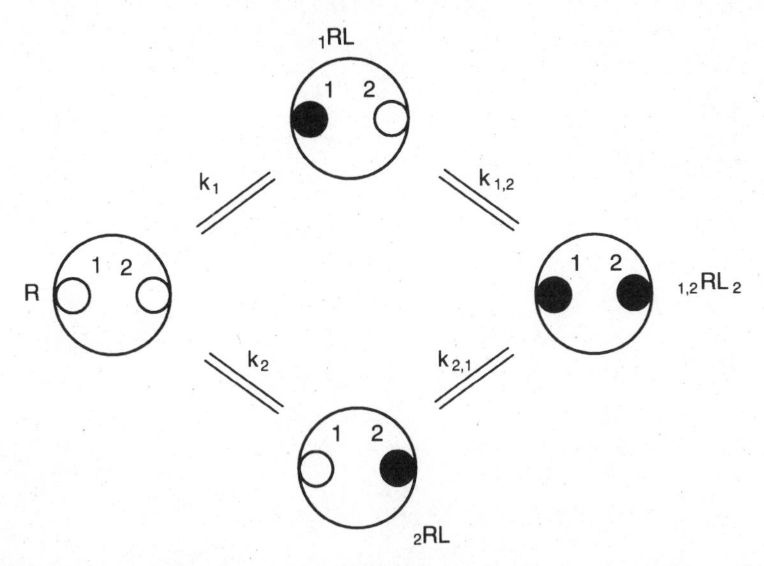

Figure 2.9 Schematic representation of a divalent receptor with two binding sites in which each site affinity changes with binding of ligand at the other site. ${}_1RL$ represents the complex with one ligand bound at site 1; ${}_2RL$, that with ligand at site 2; and ${}_{1,2}RL_2$, the species with ligands bound at both sites 1 and 2. The respective site binding constants $k_1, k_2, k_{1,2}$ and $k_{2,1}$ are indexed in an appropriate manner.

three are independent):

$$k_1 = \frac{({}_1RL)}{(R)(L)} \tag{2.22}$$

$$k_2 = \frac{({}_2RL)}{(R)(L)} \tag{2.23}$$

$$k_{1,2} = \frac{({}_{1,2}RL_2)}{({}_1RL)(L)} \tag{2.24}$$

$$k_{2,1} = \frac{({}_{1,2}RL_2)}{({}_2RL)(L)} \tag{2.25}$$

The index numbers for $k_{1,2}$ and $k_{2,1}$ specify first the site occupied and second the site being filled in the equilibrium shown. The index numbers at the lower left in ${}_{a,b}RL_m$ specify which sites are occupied by ligand in the complex with m moles of bound ligand on each receptor molecule.

When the sites have (different) *invariant* affinities, then

$$k_1 = k_{2,1} \tag{2.26a}$$

$$k_2 = k_{1,2} \tag{2.27a}$$

and one can write an equation for B, as shown in equation 2.20, with two terms. However, when the affinities vary with occupancy at other sites, then

$$k_1 \neq k_{2,1} \tag{2.26b}$$

$$k_2 \neq k_{1,2} \tag{2.27b}$$

So how should one write an equation for the moles of bound ligand B?

One might be tempted by simplistic analogy to write

$$B = \frac{k_1(L)}{1 + k_1(L)} + \frac{k_2(L)}{1 + k_2(L)} + \frac{k_{1,2}(L)}{1 + k_{1,2}(L)} + \frac{k_{2,1}(L)}{1 + k_{2,1}(L)} \tag{2.28}$$

However, even a little reflection shows that this cannot be correct. For as the ligand concentration is continuously increased, as (L)

approaches infinity, the first term in equation 2.28 approaches 1, and so does each of the succeeding terms. Hence

$$\lim_{(L) \to \infty} B(\text{predicted}) = 1 + 1 + 1 + 1 = 4 \tag{2.29}$$

But the receptor has only two sites, so the actual limit at saturation is

$$\lim_{(L) \to \infty} B(\text{actual}) = 2 \tag{2.30}$$

Obviously, equation 2.28 is incorrect. In fact it is not feasible in general to write an equation for the binding of B in terms of a series of hyperbolic terms containing each of the individual site binding constants.

2.4.2. Multivalent Systems

The number of site binding constants rises steeply as the valency of the receptor increases. For a trivalent system with different and changing site affinities, the individual constants can be designated by the schematic representation of Figure 2.10. The total number of site binding constants increases to 12, of which 7 are independent.

Furthermore, as Table 2.1 illustrates, the rise in the number of site

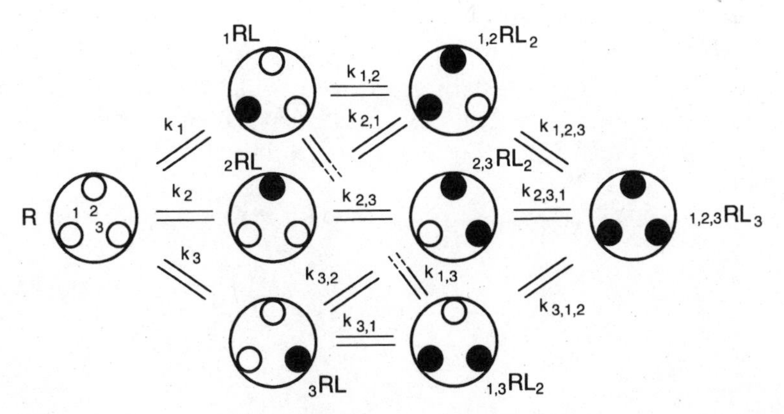

Figure 2.10 Schematic representation of site binding constants for a trivalent receptor, containing three binding sites. A total of 12 different site constants can be designated when the site affinities are different and change with extent of occupancy by ligands.

TABLE 2.1. Number of Site Binding Constants for a Receptor with *n* Binding Sites

Number of Binding Sites, n	Total Number of Site Binding Constants	Number of Independent Site Binding Constants
2	4	3
3	12	7
4	32	15
6	192	63
8	1,024	255
12	24,576	4,095

constants is even more rapid for further increases in total moles of bound ligand per mole of receptor. Thus it becomes futile to attempt to establish values for the site constants, for there are so many. In addition, many different combinations of assignments of k_j values can fit the experimental observations, even if we limit ourselves to the independent site constants. Thus, except in simple situations, it is not feasible to express a binding equation in terms of the interacting site binding constants.

REFERENCES

1. I. M. Klotz and R. M. Rosenberg, *Chemical Thermodynamics*, 5th ed., John Wiley & Sons, Inc., New York, 1994.
2. L. Michaelis and M. L. Menten, *Biochem. Z.,* **49,** 333–369 (1913).
3. B. Wolff, in J. B. S. Haldane and K. G. Stern, *Allgemeine Chemie der Enzyme,* Steinkopf, Dresden, 1932, p. 119; and J. B. S. Haldane, *Enzymes,* Longmans, Green & Co., London, 1930; reprinted by M.I.T. Press, Cambridge, Mass., 1965.
4. P. Suter and J. Rosenbusch, *J. Biol. Chem.,* **251,** 5986–5991 (1976).

CHAPTER 3

AFFINITIES: FROM A STOICHIOMETRIC PERSPECTIVE

It is also possible to express ligand binding constants in a classical format in terms of the stoichiometry of the participants in the equilibrium, without any consideration of the sites involved. In this thermodynamic formulation, we shall also proceed from the simplest case to the general.

3.1. DIVALENT RECEPTOR

The uptake of ligand L by receptor R may be represented by the two stoichiometric steps:

$$R + L = RL_1 \tag{3.1}$$

$$RL_1 + L = RL_2 \tag{3.2}$$

For each we write an equilibrium constant capital K_i (to be distinguished from the site constant k_j)

$$K_1 = \frac{(RL_1)}{(R)(L)} \tag{3.3}$$

$$K_2 = \frac{(RL_2)}{(RL_1)(L)} \tag{3.4}$$

Each of these equations can be rearranged to focus on the concentrations of the respective liganded species:

$$(RL_1) = K_1(R)(L) \tag{3.5}$$

$$(RL_2) = K_2(RL_1)(L) = K_1K_2(R)(L)^2 \tag{3.6}$$

Now we are prepared to derive an equation for moles bound ligand B:

$$B = \frac{\text{moles bound ligand}}{\text{moles total receptor}} = \frac{(RL_1) + 2(RL_2)}{(R) + (RL_1) + (RL_2)} \tag{3.7}$$

The numerator contains the concentration of (RL_1), which contributes 1 mole of bound ligand, and $2(RL_2)$ because there are 2 moles of bound L in each RL_2. In the denominator, each receptor stoichiometric species—R, RL_1, and RL_2—contributes 1 mole of receptor to the total. Inserting the relations of equations 3.5 and 3.6 into equation 3.7 and canceling out the common factor (R) that appears in numerator and denominator, we obtain the stoichiometric binding equation:

$$B = \frac{K_1(L) + 2K_1K_2(L)^2}{1 + K_1(L) + K_1K_2(L)^2} \tag{3.8}$$

The dependence of B on (L) is governed by the two stoichiometric binding constants K_1 and K_2. Their numerical values can be determined from a given set of experimental data for B and (L). Two typical results are (1,2)

Human serum transferrin–phosphate	$K_1 = 1.5 \times 10^4$; $K_2 = 0.2 \times 10^4$
Calbindin–calcium	$K_1 = 2.2 \times 10^8$; $K_2 = 3.7 \times 10^8$

3.2. TETRAVALENT RECEPTOR

Before outlining the general treatment let us present the format for a tetravalent receptor, since hemoglobin, Hb, plays such an important role not only in human physiology but also as a model for probing various features of ligand–receptor complexes.

For a tetravalent receptor, the stepwise stoichiometric equilibria and the associated equations for the stoichiometric equilibrium constants and species concentrations can be presented concisely as follows:

$$R + L = RL_1; \quad K_1 = \frac{(RL_1)}{(R)(L)}; \qquad (RL_1) = K_1(R)(L) \tag{3.9}$$

$$RL_1 + L = RL_2; \quad K_2 = \frac{(RL_2)}{(RL_1)(L)}; \quad (RL_2) = K_1K_2(R)(L)^2 \tag{3.10}$$

$$RL_2 + L = RL_3; \quad K_3 = \frac{(RL_3)}{(RL_2)(L)}; \quad (RL_3) = K_1K_2K_3(R)(L)^3 \tag{3.11}$$

$$RL_3 + L = RL_4; \quad K_4 = \frac{(RL_4)}{(RL_3)(L)}; \quad (RL_4) = K_1K_2K_3K_4(R)(L)^4 \tag{3.12}$$

Corresponding to equation 3.7 for a divalent system, we find

$$B = \frac{(RL_1) + 2(RL_2) + 3(RL_3) + 4(RL_4)}{(R) + (RL_1) + (RL_2) + (RL_3) + (RL_4)} \tag{3.13}$$

In the numerator, the numerical coefficients 1, 2, 3, and 4 present the number of bound ligands contributed by the respective species RL_i. In the denominator, each coefficient is 1 since each species RL_i contributes a single receptor R. Substituting the relations for (RL_i) given in equations 3.9 to 3.12 we can obtain

$$B = \frac{K_1(L) + 2K_1K_2(L)^2 + 3K_1K_2K_3(L)^3 + 4K_1K_2K_3K_4(L)^4}{1 + K_1(L) + K_1K_2(L)^2 + K_1K_2K_3(L)^3 + K_1K_2K_3K_4(L)^4} \tag{3.14}$$

For hemoglobin–oxygen equilibria, careful experimental studies have been carried out for many decades and stoichiometric constants have been evaluated by a variety of statistical best-fit methods. A representative set of results is the following (3):

$$K_1 = 0.0188$$
$$K_2 = 0.0566$$
$$K_3 = 0.407$$
$$K_4 = 4.28$$

3.3. MULTIVALENT RECEPTOR

With the background of the special cases of divalent and tetravalent complexes, we can readily formulate the appropriate relationships for the general case of a multivalent receptor:

$$R + L = RL_1; \quad K_1 = \frac{(RL_1)}{(R)(L)}; \qquad (RL_1) = K_1(R)(L) \tag{3.15}$$

$$RL_1 + L = RL_2; \quad K_2 = \frac{(RL_2)}{(RL_1)(L)}; \qquad (RL_2) = K_1K_2(R)(L)^2 \tag{3.16}$$

$$\vdots \qquad\qquad\qquad\qquad \vdots$$

$$RL_{i-1} + L = RL_i; \quad K_i = \frac{(RL_i)}{(RL_{i-1})(L)}; \quad (RL_i) = (K_1K_2 \cdots K_i)(R)(L)^i \tag{3.17}$$

$$\vdots \qquad\qquad\qquad\qquad \vdots$$

$$RL_{n-1} + L = RL_n; \quad K_n = \frac{(RL_n)}{(RL_{n-1})(L)}; \quad (RL_n) = (K_1 \cdots K_n)(R)(L)^n \tag{3.18}$$

We terminate the series of steps at saturation of the receptor; n represents the moles of bound ligand per mole receptor at saturation. We recognize that n is also the number of ligand-binding sites on a receptor molecule.

Generalizing the steps taken in the divalent and tetravalent systems, we can express the equation for B for a multivalent system by

$$B = \frac{(RL_1) + 2(RL_2) + \cdots + i(RL_i) + \cdots + n(RL_n)}{(R) + (RL_1) + (RL_2) + \cdots + (RL_i) + \cdots + (RL_n)} \tag{3.19}$$

With the necessary substitutions, obtained from equations 3.15 to 3.18, we arrive at the general expression

$$B = \frac{K_1(L) + 2K_1K_2(L)^2 + \cdots + i\left(\prod_{l=1}^{i} K_l\right)(L)^i + \cdots + n\left(\prod_{l=1}^{n} K_l\right)(L)^n}{1 + K_1(L) + K_1K_2(L)^2 + \cdots + \left(\prod_{l=1}^{i} K_l\right)(L)^i + \cdots + \left(\prod_{l=1}^{n} K_l\right)(L)^n} \tag{3.20}$$

The successive stoichiometric constants can be evaluated by procedures for best-fitting of equation 3.20 to experimental data. For example, for the binding of laurate ions by human serum albumin (4), iterative least squares fitting to 220 experimental points provided K_i values up to $i = 10$. The stoichiometric constants giving the best-fit approximation are the following:

Stoichiometric Step	Stoichiometric Binding Constants (M^{-1})
1	8.31×10^6
2	1.57×10^6
3	2.88×10^5
4	6.18×10^5
5	1.07×10^2
6	1.98×10^8
7	9.25×10^1
8	1.43×10^7
9	1.86×10^3
10	1.94×10^4

It should be stressed that the termination of entries at K_{10} does not imply that the receptor serum albumin is saturated with ligand at $n = 10$. The actual experimental data give no indication of a saturation plateau. Since experimental data for B values appreciably above 9 mol/mol receptor are not available, K_i values above $i = 10$ cannot be determined.

REFERENCES

1. W. R. Harris, *Biochemistry*, **24**, 7412–7418 (1985).
2. S. Linse, P. Broden, T. Drakenberg, E. Thulin, P. Sellers, K. Elmden, T. Grundström, and S. Forsén, *Biochemistry*, **26**, 6723–6735 (1987).
3. K. Imai, *Meth. Enzymol.*, **232**, 559–655 (1994).
4. A. O. Pedersen, B. Hust, S. Andersen, F. Neelsen, and R. Brodersen, *Eur. J. Biochem.*, **154**, 545–552 (1986).

CHAPTER 4

AFFINITIES: FROM A GHOST-SITE PERSPECTIVE

4.1. GHOST-SITE BINDING CONSTANTS

The stoichiometric binding equation 3.20 can be transformed into an alternative expression that offers a useful perspective into ligand–receptor affinities.

Inspection of equation 3.20 reveals that the denominator Z is a polynomial of degree n. Recalling the fundamental theorem of algebra that a polynomial of degree n has n (and only n) roots, we may write

$$
\begin{aligned}
Z &= 1 + K_1 L + K_1 K_2 L^2 + \cdots + \left(\prod_{l=1}^{n} K_l \right) L^n \\
&= (\Pi K_l)[(L - a_1)(L - a_2) \cdots (L - a_n)]
\end{aligned}
\tag{4.1}
$$

where $a_1, a_2, \cdots a_n$ are the roots of Z.

In addition, if we differentiate the denominator of equation 3.20, term by term, with respect to L, we obtain

$$
\frac{dZ}{dL} = K_1 + 2K_1 K_2 L + \cdots + n \left(\prod_{l=1}^{n} K_l \right) L^{n-1}
\tag{4.2}
$$

Hence

$$L\,\frac{dZ}{dL} = K_1L + 2K_1K_2L^2 + \cdots + n\left(\prod_{l=1}^{n} K_l\right)L^n \qquad (4.3)$$

which is the numerator of equation 3.20. Thus we conclude that

$$\frac{L\,\dfrac{dZ}{dL}}{Z} = \frac{\text{numerator of equation 3.20}}{\text{denominator of equation 3.20}} = B \qquad (4.4)$$

Applying this relation to the expression for Z in equation 4.1, we write

$$\begin{aligned} B &= L\,\frac{1}{Z}\cdot\frac{dZ}{dL} = L\,\frac{d\ln Z}{dL} \\ &= L\left[\frac{d\ln(\Pi K_l)}{dL} + \frac{d\ln(L-a_1)}{dL} + \frac{d\ln(L-a_2)}{dL} + \cdots\right] \\ &= \frac{L}{L-a_1} + \frac{L}{L-a_2} + \cdots + \frac{L}{L-a_n} \end{aligned} \qquad (4.5)$$

This can be converted to

$$B = \frac{\left(-\dfrac{1}{a_1}\right)L}{1+\left(-\dfrac{1}{a_1}\right)L} + \frac{\left(-\dfrac{1}{a_2}\right)L}{1+\left(-\dfrac{1}{a_2}\right)L} + \cdots + \frac{\left(-\dfrac{1}{a_n}\right)L}{1+\left(-\dfrac{1}{a_n}\right)L} \qquad (4.6)$$

For a specific polynomial Z, the roots are constants. Hence we can replace each $(-1/a_l)$ in equation 4.6 by a corresponding constant $K_\alpha, K_\beta \cdots K_\nu$ and obtain

$$B = \frac{K_\alpha L}{1+K_\alpha L} + \frac{K_\beta L}{1+K_\beta L} + \cdots + \frac{K_\nu L}{1+K_\nu L} \qquad (4.7)$$

(with $\nu = n$).

This equation is rigorously equivalent to equation 3.20, which was derived from purely thermodynamic stoichiometric considerations. Nevertheless, the constants $K_\alpha \cdots K_\nu$ do *not* correspond to the stoichiometric binding constants $K_1 \cdots K_n$. Equation 4.7 was derived without taking any cognizance of the affinity characteristics of binding sites on the receptor or even of their existence. Nevertheless, equation 4.7 resembles the site-binding equation 2.20, which is valid only for sites with different but invariant affinities. However, in contrast to equation 2.20, in the derivation of equation 4.7 we did not introduce any assumptions about the affinities of the sites. Therefore, equation 4.7 is valid even when site affinities *change* with the extent of occupancy by ligands, whereas equation 2.20 or 2.28 collapses. Thus it is apparent that *in general* the constants $K_\alpha \cdots K_\nu$ do *not* correspond to the site-binding constants $k_1, \cdots k_n$.

For illustration, we can compare the values of K_α and K_β with those for K_1 and K_2 for the divalent receptors listed earlier:

Human serum transferrin–phosphate	$K_1 = 1.5 \times 10^4; K_\alpha = 1.3 \times 10^4$
	$K_2 = 0.2 \times 10^4; K_\beta = 0.25 \times 10^4$
Calbindin–calcium	$K_1 = 2.2 \times 10^8; K_\alpha = 2.9 \times 10^8 e^{0.38\pi i}$
	$K_2 = 3.7 \times 10^8; K_\beta = 2.9 \times 10^8 e^{-0.38\pi i}$

(The i in the exponential factors is the symbol for the imaginary number $\sqrt{-1}$.) We cannot list any comparison values for the site constants for these divalent systems, since at this stage we do not know whether the affinities of the binding sites are invariant or change with occupancy by ligand (actually in these examples they change with occupancy by ligand).

A striking feature of K_α and K_β is that they may assume complex, imaginary values, as is shown, for example, for calbindin–calcium. This example is not extraordinary. As we shall see later, complex numbers for these binding constants appear in many cases.

Since $K_\alpha \cdots K_\nu$ can have imaginary values, it seems apt to denote them *virtual binding constants*, or *ghost-site binding constants*.

4.2. ALTERNATIVE GHOST-SITE BINDING EQUATION

When the ghost-site constants $K_\alpha \cdots K_\nu$ are complex numbers, they are coupled in a way that leads to an alternative algebraic expression for B. To illustrate, let us consider a bivalent receptor, which becomes saturated when two moles of ligand are bound.

Since complex roots must appear as conjugate pairs, we can write

$$K_\alpha = a + bi; \; K_\beta = a - bi \tag{4.8}$$

where $i = \sqrt{-1}$ and the coefficients a and b are real numbers. In exponential form

$$K_\alpha = Ae^{i\theta}; \; K_\beta = Ae^{-i\theta} \tag{4.9}$$

where $A = \sqrt{a^2 + b^2}$ and $\theta = \arctan(b/a)$. Insertion of the expressions of equation 4.9 into equation 4.7 leads to

$$B = \frac{Ae^{i\theta}L}{1 + Ae^{i\theta}L} + \frac{Ae^{-i\theta}L}{1 + Ae^{-i\theta}L} \tag{4.10}$$

Combining the two terms into one over a common denominator, we find

$$B = \frac{A(e^{i\theta} + e^{-i\theta})L + 2A^2L^2}{1 + A(e^{i\theta} + e^{-i\theta})L + A^2L^2} \tag{4.11}$$

Recalling the relation

$$e^{i\theta} + e^{-i\theta} = 2\cos\theta \tag{4.12}$$

we obtain

$$B = \frac{A(2\cos\theta)L + 2A^2L^2}{1 + A(2\cos\theta)L + A^2L^2} \tag{4.13}$$

This equation for B is an alternative to equation 4.7 with two terms. For the divalent system calbindin–calcium, for example,

$$B = \frac{2.9 \times 10^8(2 \times 0.37)L + 2(2.9)^2 \times 10^{16}L^2}{1 + 2.9 \times 10^8(2 \times 0.37)L + (2.9)^2 \times 10^{16}L^2} \tag{4.14}$$

Turning to a trivalent system, in which saturation of the receptor is attained with three bound ligands, we recognize that the ghost-site binding equation 4.7 will have three terms. One pair of the constants $K_\alpha, K_\beta, K_\gamma$ can be complex conjugates, the other (assigned to K_γ) must be real. In this situation, the equation corresponding to 4.13 becomes

$$B = \frac{A(2\cos\theta)L + 2A^2L^2}{1 + A(2\cos\theta)L + A^2L^2} + \frac{K_\gamma L}{1 + K_\gamma L} \tag{4.15}$$

For a tetravalent system there can be one or two pairs of complex conjugate values for the ghost constants K_ω. For the latter case, the alternative equation for binding becomes

$$B = \sum_{l=1}^{2(=4/2)} \frac{A_l(2\cos\theta_l)L + 2A_l^2L^2}{1 + A_l(2\cos\theta_l)L^2 + A_l^2L^2} \tag{4.16}$$

In turn, if in any specific system only one pair of complex roots appears, then

$$B = \frac{A(2\cos\theta)L + 2A^2L^2}{1 + A(2\cos\theta)L + A^2L^2} + \sum_{K_\omega = K_\gamma}^{K_\delta} \frac{K_\omega L}{1 + K_\omega L} \tag{4.17}$$

In the general case where saturation of receptor is reached with n ($= \nu$) moles of bound ligand (and n is an even number),

$$B = \sum_{l=1}^{s \le n/2} \frac{A_l(2\cos\theta_l)L + 2A_l^2L^2}{1 + A_l(2\cos\theta_l)L + A_l^2L^2} + \sum_{\omega_1}^{\omega_{n-2s}} \frac{K_\omega L}{1 + K_\omega L} \tag{4.18}$$

CHAPTER 5

FACTS AND FANTASIES FROM GRAPHICAL ANALYSES

5.1. DIRECT SCALES

The simplest plot to prepare to present binding data is one of B versus (L) (see Fig. 2.5). If one can unequivocally approach a plateau for B, then the moles of ligand bound at saturation n can be established. Thereafter, by analogy with the simple situation presented in Figure 2.5, one might be inclined to estimate an affinity from the value of $1/(L)$ at which $B = n/2$.

In actuality, except in a few special cases, this practice is treacherous and misleading. The pitfalls encountered are strikingly illustrated by the example presented in Figure 5.1. This graph displays data from an exceptionally extensive and detailed study (encompassing more than 200 data points) of the binding of laurate (dodecanoate) ligands by human serum albumin (1,2). Figure 5.1a shows the binding isotherm for the collection of experiments up to an (L) value of 16 μM. With such a large collection of experimental points, the viewer would feel confident that saturation must be attained at $n = 7$. In the actual study, additional experiments extending the range of (L) sixfold were carried out at higher ligand concentrations up to about 50 μM (Fig. 5.1b). Inspection of Figure 5.1b shows that $n = 7$ has been passed, and an answer of $n = 8$ seems eminently reasonable. Once again, however, additional binding experiments that extend the range of (L) an additional sixfold, to 300 μM (Fig. 5.1c) demonstrate that the plateau n

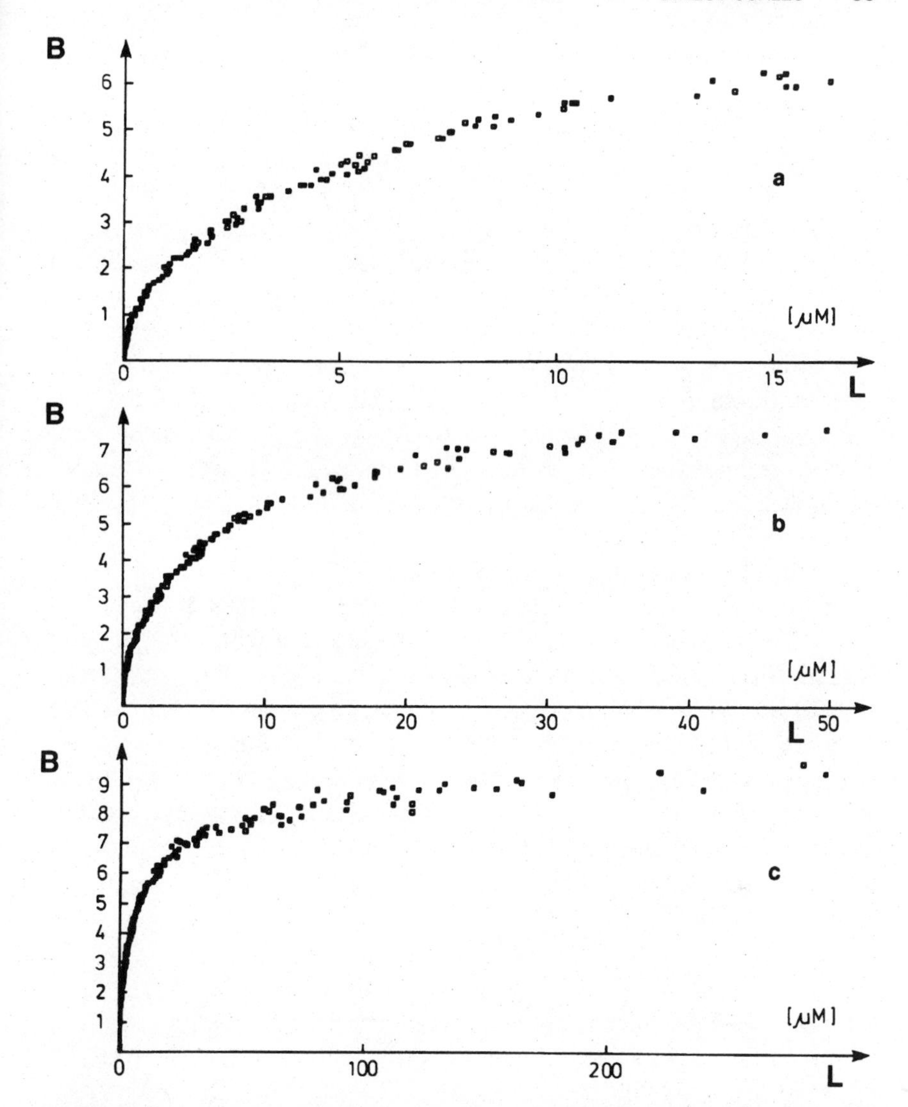

Figure 5.1 Linear scale plots of the binding of laurate ion by human serum albumin, showing three different ranges of concentration.

estimated from Figure 5.1b is also in error. At this stage, one might become wary and reluctant to state that the plateau is attained even at $n = 10$. In fact, as we shall see later, from an alternative graphical presentation it becomes obvious that despite the collection of 220 data points over a range of 10^6-fold in concentration of free ligand (L) saturation of this receptor has not been attained.

Another disadvantage of a B versus (L) graph, clearly evident in Figure 5.1, is the enormous compression of the data at low values of (L). This defect is compounded the greater the range of concentrations covered in the study. In essence, one loses the ability to read individual points.

There are no other published studies with such an extensive and detailed range of experimental observations. Most present fewer than 10 points, over less than a 10-fold range of (L), and many are even more constricted in scope. In general, a direct plot of B versus (L) leads to erroneous conclusions about the total number of binding sites and is not convenient for further insights into binding affinities.

Nevertheless, in a few cases, a B–(L) plot can establish the number of binding sites with confidence. This occurs when the receptor's affinity increases with increasing binding of ligand, when binding is "cooperative." An example of such behavior is illustrated in Figure 5.2. For the binding of leucine by α-isopropylmalate synthase (3), the two successive stoichiometric constants are $K_1 = 0.48 \times 10^5$ and $K_2 = 2.5 \times 10^5$, clearly reflecting the increasing affinity with occupancy by ligand. B rises rapidly to a clear plateau, a horizontal line over a range of 20-fold in (L). Even in this case, however, as Figure 5.2 shows, the data are highly compressed in the lower binding range.

A classical example of unequivocal saturation of receptor is the hemoglobin–oxygen equilibrium; a prototypical curve is illustrated in Figure 5.3. That oxygen affinity of hemoglobin increases with occu-

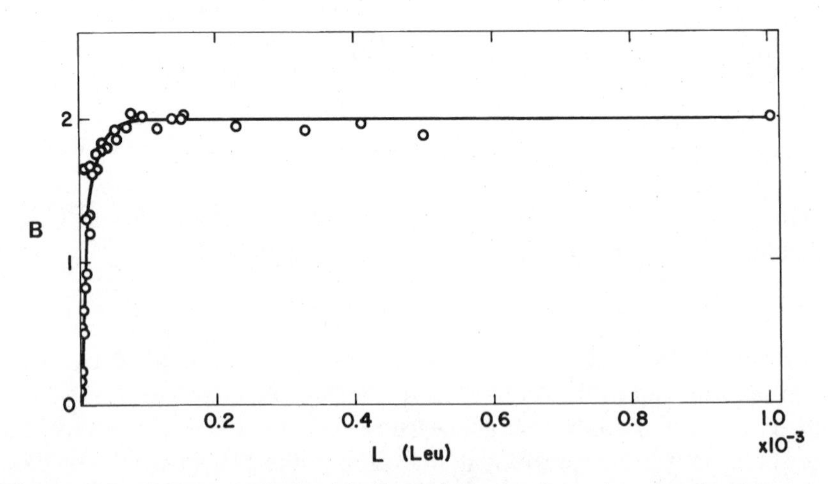

Figure 5.2 Direct graph for binding of leucine by α-isopropylmalate synthase using linear scales for B and (L).

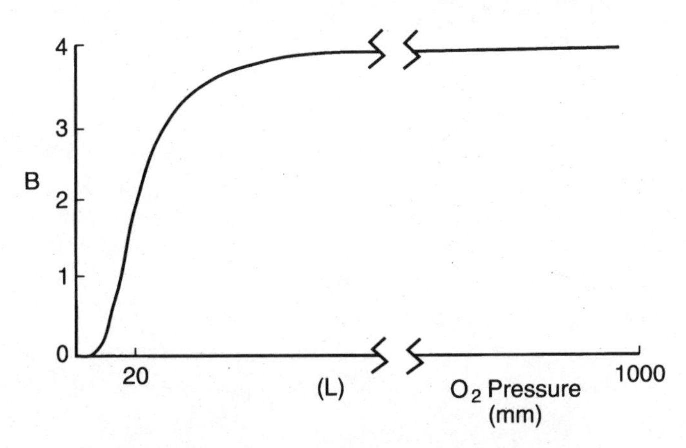

Figure 5.3 Uptake of oxygen by hemoglobin.

pancy of sites on the receptor is shown by the values of the stoichiometric binding constants cited in Chapter 3: $K_1 = 0.0188$, $K_2 = 0.0566$, $K_3 = 0.407$, and $K_4 = 4.28$. In hemoglobin–oxygen equilibria, the investigator focuses primarily on the uptake at low oxygen tensions, i.e., on data at the low end of (L), for this is the range of physiological interest. To circumvent compression of the binding curve at this low end, the data at very high oxygen tensions are usually omitted, for it has been unequivocally established for almost a century that at saturation 4 moles of O_2 are bound by (human and other mammalian) hemoglobin.

5.2. LINEAR TRANSFORM GRAPHS

In the overwhelming majority of ligand–receptor binding studies, it is not really possible to see the moles bound at saturation of receptor, i.e., the number of sites, from a graph of B versus (L). Consequently, experimenters have turned to graphs of one of the linear transforms of equation 2.14, among which the Scatchard plot, a single reciprocal graph, has been the most widely used. It is rarely appreciated that such a plot introduces a severe compression of $B/(L)$ data as (L) increases progressively. As a result, one can be led to conclusions that are untenable.

When data are plotted on a $B/(L)$ versus (L) graph, there is a strong, sometimes irresistable, temptation to fit them to a straight line, either by eye or by least square methods. Figure 5.4a summarizes the experimental observations (4) for the binding of diazepam by benzodiazepine receptors. Individual points have been fitted to a straight line. To the eye

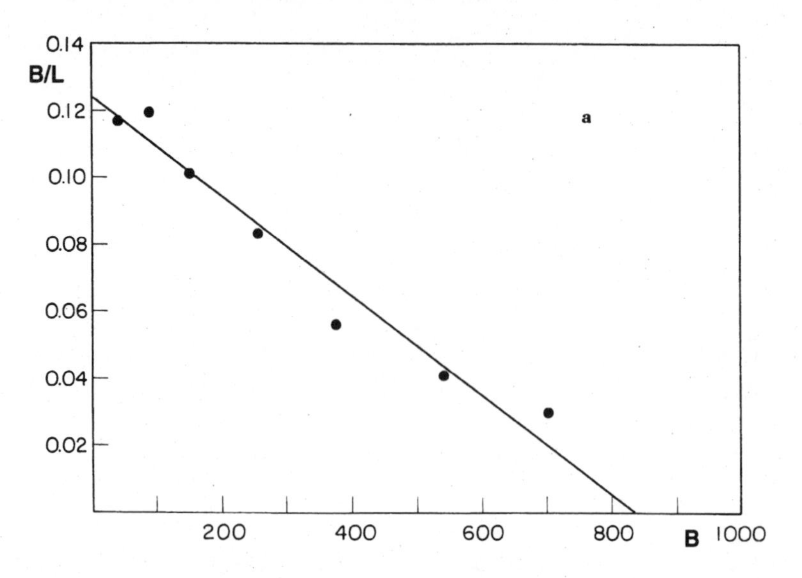

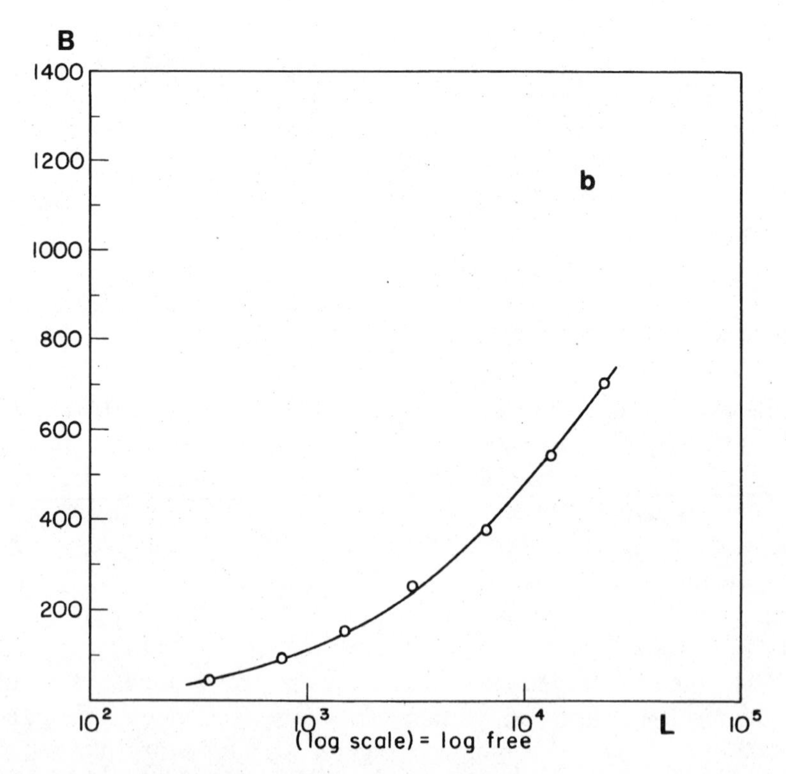

Figure 5.4 Binding of diazepam by benzodiazepine receptors. a, Scatchard graph. b, Semilogarithmic plot. Data from Ref. 4.

that seems reasonable, and such a judgment is apparently verified by a good correlation coefficient (0.97) in a statistical analysis. The linearity implies that all the sites are identical in affinity and invariant. In addition, the total number of receptor sites, 830, was determined from an extrapolation of the least squares line to the intercept on the *B* axis (4), in accord with equation 2.19.

However, if the same data points are plotted on a graph of *B* versus log *L* (Fig. 5.4b), it becomes obvious that there is not even a hint of saturation at 830. Furthermore, since the graph in Figure 5.4a is purported to be a straight line, the one in Figure 5.4b should be an ideal S-shaped curve (like that in Fig. 2.6), which at its inflexion point would have a value of *B* halfway to the saturation plateau (5). Even if we arbitrarily assume that the highest observed experimental value of *B*, about 700, is at the inflexion point in Figure 5.4b, the saturation plateau would be double this value, i.e., 1400; actually it may be even higher. Consequently, in the Scatchard plot, the intercept on the *B* axis must be far to the right of that shown. Furthermore, the experimental points should not be fitted to a straight line; the correlation coefficient of 0.97 for a straight-line fit is misleading.

An example of a preposterous fit of binding data to a straight line in a Scatchard plot is shown in Figure 5.5. In this case, the proposers of a

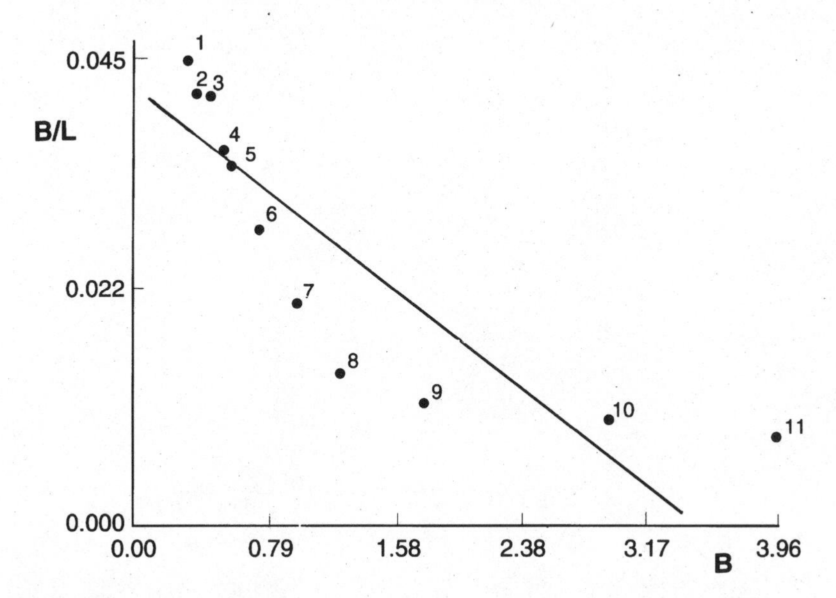

Figure 5.5 A computer-generated linear fit in a Scatchard plot. The individual data points have been labeled ordinally.

linear fit appear to be oblivious to the actual data and seem to believe that because a linear fit was obtained with a computer program, human subjective judgment has been circumvented (6).

Even when the nonlinearity in a B/L versus B plot cannot be overlooked, the graph may entice one into untenable conclusions. A serious error that is pervasive throughout the literature is the drawing of two straight lines through a set of data, such as those presented in Figure 5.6 for the binding of calcium ions by a mitochondrial transport protein (7). The authors of such plots evidently believe that if there are two classes of binding sites, with different but identical, invariant affinities within each class, the high affinity class data should fit on one of the straight lines shown and the weaker class points should be placed on another straight line. This misconception arises from a misunderstanding of the algebraic features of the binding equations.

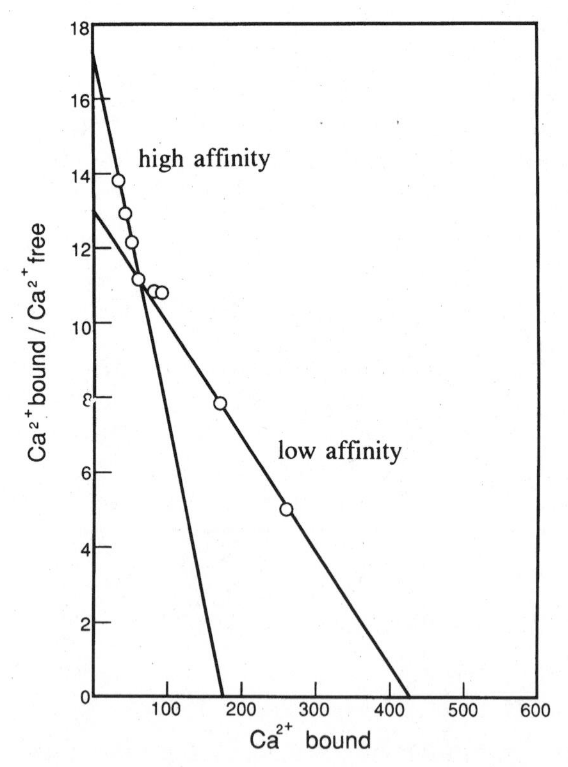

Figure 5.6 Scatchard graph for the binding of calcium ions by a transport protein from mitochondrial membranes. The inappropriate assignments of "high affinity" and "low affinity" sites were made in the original paper (7).

For a single set of identical invariant sites,

$$B = \frac{nk(L)}{1 + k(L)} \tag{2.14}$$

which can be transformed to the linear relation

$$B/(L) = nk - kB \tag{2.19}$$

From this equation, it follows that a plot of $B/(L)$ versus B should be linear for a *single* set of identical invariant sites. For two classes of sites, with (different but) identical invariant affinities within each class,

$$B = \frac{n_1 k_1 (L)}{1 + k_1 (L)} + \frac{n_2 k_2 (L)}{1 + k_2 (L)} \tag{2.21}$$

where the subscripts denote the respective number of sites and the binding constants of the two individual classes of sites. Equation 2.21 *can* be deconstructed into the two lines shown in Figure 5.6 *only if* one of the two terms is negligibly small compared to the other. In any event, it is simply not true that the high affinity line in Figure 5.6 has a slope equal to $-k_1$ and an intercept equal to n_1 and that the low affinity line is appropriate for characterizing the second class of binding sites (see Appendix A2).

The actual values of the intercepts and slopes in a Scatchard graph for multiple classes of identical invariant sites are much more complicated (8). The complexity is evident even for the simplest such system, just two independent binding sites (Fig. 5.7). The first intercept on the B axis is *not* equal to 1, i.e., n_1, although the second equals the *total* number of sites 2 not n_2. The respective intercepts on the $B(L)$ axis are *not* equal to k_1 or k_2 but are complicated functions of the site parameters. So are also the limiting slopes of the lines to the curves as (L) goes to zero and to very high concentrations, respectively.

General equations for the intercepts and slopes when there are multiple classes of identical invariant binding sites are derived in Appendix A2.

The type of error exemplified by Figure 5.6 is pervasive in the literature. A collection of such incorrect deconvolutions of curved Scatchard plots has been assembled (9).

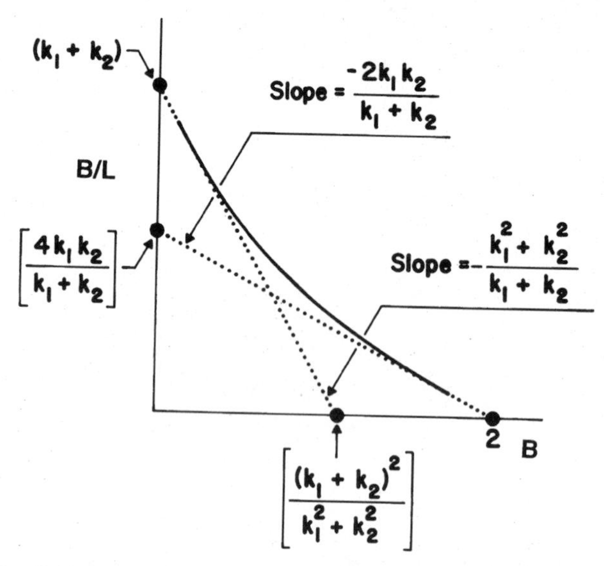

Figure 5.7 Schematic curves for a system of two independent sites. The actual intercepts and limiting slopes are denoted.

Even when it is recognized that a curved Scatchard graph should not be deconvoluted into two asymptotic straight lines, it is rarely appreciated how unreliable are attempts to determine an intercept on the B-axis, the total number of sites. Experimental points become extremely compressed as $B/(L)$ approaches zero, as it must at the intercept on the B axis. This situation is illustrated in Figure 5.8, which presents data for the binding of carbamoyl phosphate by aspartate transcarbamoylase (10). If the numerical values are deleted from the coordinate axes, it becomes evident that it is not possible to establish a point for the intercept along the B axis, even if it existed. When numerical values are shown explicitly, one is tempted to draw a line through the enormously compressed points at low $B/(L)$ values (in the region where (L) increases from 100 to 1000 μM to ∞) so that there is an intercept at 6 moles of bound ligand per enzyme receptor, the number of binding sites known from extensive structural information.

Curves such as that shown in Figure 5.8 can often be fitted by an equation for two classes of binding sites with (different but) identical, invariant affinities within each class (11),

$$B = \frac{n_1 k_1(L)}{1 + k_1(L)} + \frac{n_2 k_2(L)}{1 + k_2(L)} \tag{3.21}$$

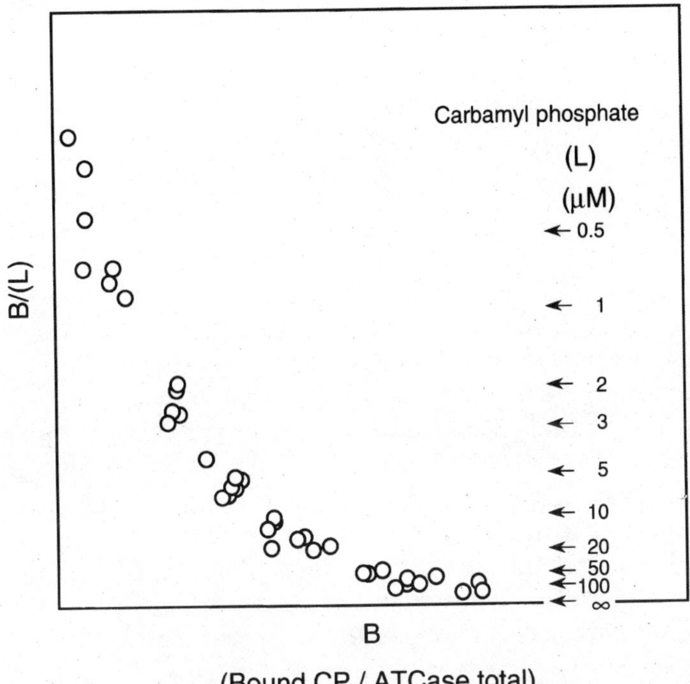

Figure 5.8 Scatchard graph of data for binding of carbamyl phosphate (CP) by aspartate transcarbamoylase (ATCase). To emphasize compression in the ordinate axis at the left, corresponding values of (*L*) are shown along the right. Numerical values along the axes have been deleted so that readers will not be tempted to pick a desired intercept.

On this basis it is often concluded that the receptor must have two classes of sites: n_1 in the first class and n_2 in the second. For example, for carbamoyl phosphate binding by aspartate transcarbamoylase it has been concluded that $n_1 = 2$ and $n_2 = 4$ (11). However, this conclusion is untenable; it is known from structural data that this enzyme is actually constituted of six identical protomers.

How does it happen that for an enzyme composed of six identical protomers (each with one binding site), the concentration dependence of the uptake of ligand can be fitted by equation 3.21, one for two classes of sites? In Chapter 4 it was shown that a formally similar binding equation is obtained without arbitrarily assuming classes of sites. From purely stoichiometric considerations, coupled with the fundamental theorem of algebra, we obtained the binding equation (4.7), which for a two-term situation becomes

$$B = \frac{K_\alpha(L)}{1 + K_\alpha(L)} + \frac{K_\beta(L)}{1 + K_\beta(L)} \tag{5.1}$$

But K_α and K_β are (in general) *not* the site binding equilibrium constants k_1 and k_2. If one postulates specific site models for a receptor, one can obtain explicit relationships between K_α and the site constants and between K_β and the site constants, as will be demonstrated later, but these are only pertinent to the specific model.

To summarize, concave curvature in a Scatchard plot does not imply that the receptor has two or more classes of sites each with identical affinities. In fact, all the sites may be identical at the outset; but if the affinities change with increasing occupancy by ligand, the binding equation 5.1 is still valid and the Scatchard graph will be curved.

Finally, in connection with Scatchard plots, it may be illuminating to see what the extensive data in Figure 5.1 look like in a graph of $B/(L)$ versus B (Fig. 5.9). As $B/(L)$ approaches zero, i.e., as the free ligand concentration rises to increasingly larger values, the curve approaches zero slope along the B axis. It is clearly impossible to pick an intercept on the B axis, and hence impossible to decide on the value of n, the total number of sites on the albumin receptor. Furthermore,

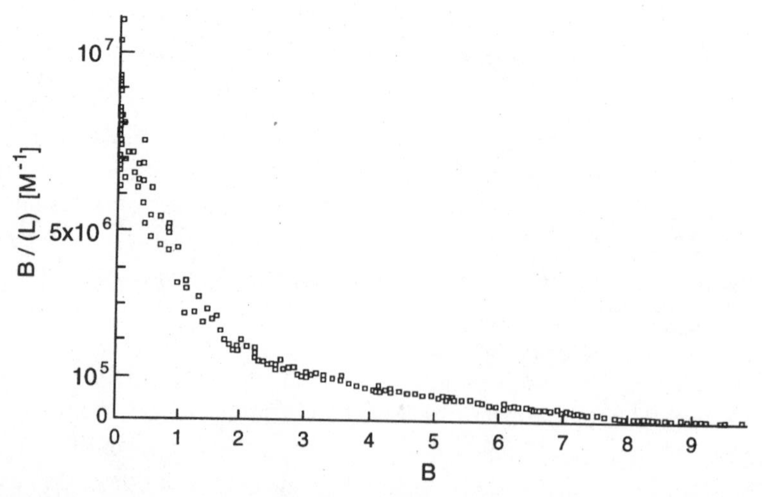

Figure 5.9 Scatchard plot of the binding of laurate ion by human serum albumin, using the same data shown in Figure 5.1.

the curve cannot be fitted by an equation with two hyperbolic terms, such as equation 2.21 or equation 5.1, but needs many more terms. As we shall see later, in laurate–albumin complexes, affinity changes with increasing binding of ligand, so that a binding equation in terms of site binding constants cannot be applied. In contrast, a description in terms of stoichiometric binding constants, equation 3.20 does give a precise representation of the binding of laurate by albumin.

5.3. SEMILOGARITHMIC GRAPHS

The most useful graphical mode of presentation of ligand–receptor binding data is a plot of B versus log (L). Its advantages will be evident from three examples.

Figure 5.10 shows a semilogarithmic transform of the data in Figure 5.2 for leucine–isopropylmalate synthase complexes and confirms that saturation of receptor occurs at $n = 2$. In addition, the binding data are spread out evenly over the large range of ligand concentration, not highly compressed at the low end.

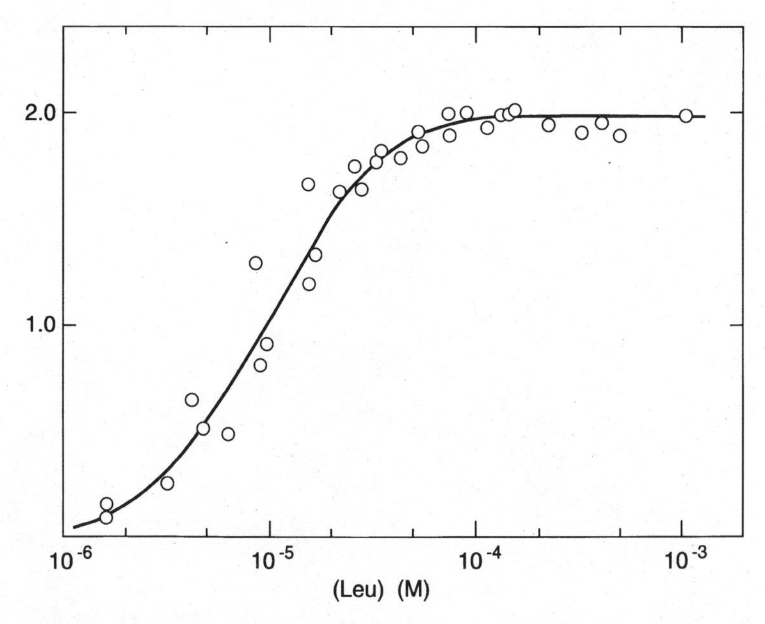

Figure 5.10 Semilogarithmic representation of the binding of leucine by isopropylmalate synthase, using the same data shown in Figure 5.2.

Figure 5.11 presents the same binding data for carbamoyl phosphate and aspartate transcarbamoylase shown in Figure 5.8 and confirms the assessment that an intercept on the B axis cannot be determined in the latter graph. In Figure 5.11, it is obvious that no plateau has been reached at high ligand concentrations. In fact, reminiscent of Figure 5.4b, the graph does not even show an inflexion point at B values as high as 5 to 6, so saturation by ligand must occur at much higher values.

The semilogarithmic graphical transform of the data shown in Figure 5.1 is presented in Figure 5.12, which makes it obvious why the total number of binding sites cannot be determined for laurate–albumin interactions. Despite the extensive binding data, covering a range of a millionfold in free ligand concentration, a plateau has not been reached at high (L). This figure also spreads out the binding data uniformly over a 10^6-fold range in free ligand concentration.

In the rare circumstance that the binding sites are all identical and invariant in affinity, a B versus log (L) curve will be symmetrical about an inflexion point, as shown in Figure 2.6. In such a case, one can eval-

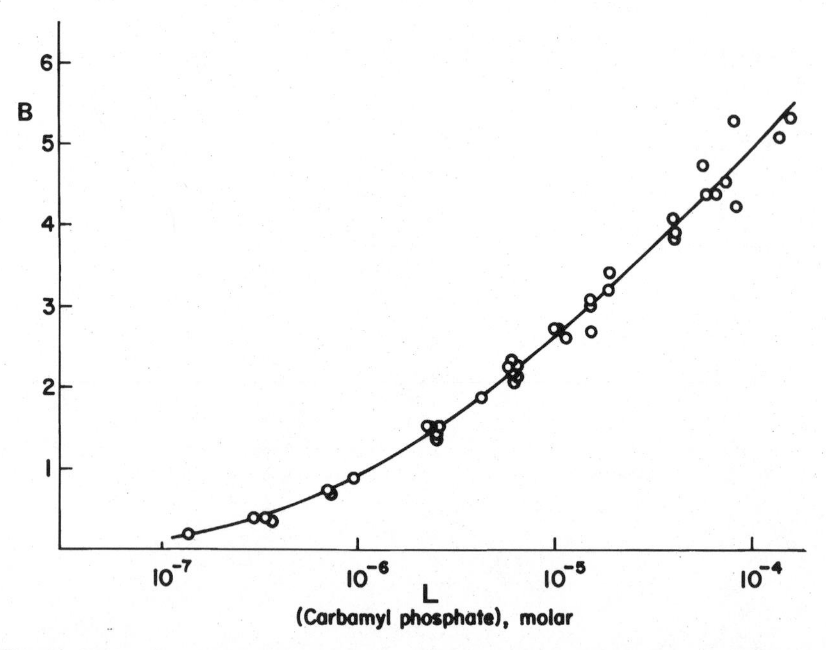

Figure 5.11 Semilogarithmic graph of the binding of carbamoyl phosphate by aspartate transcarbamoylase, using the same data shown in Figure 5.8.

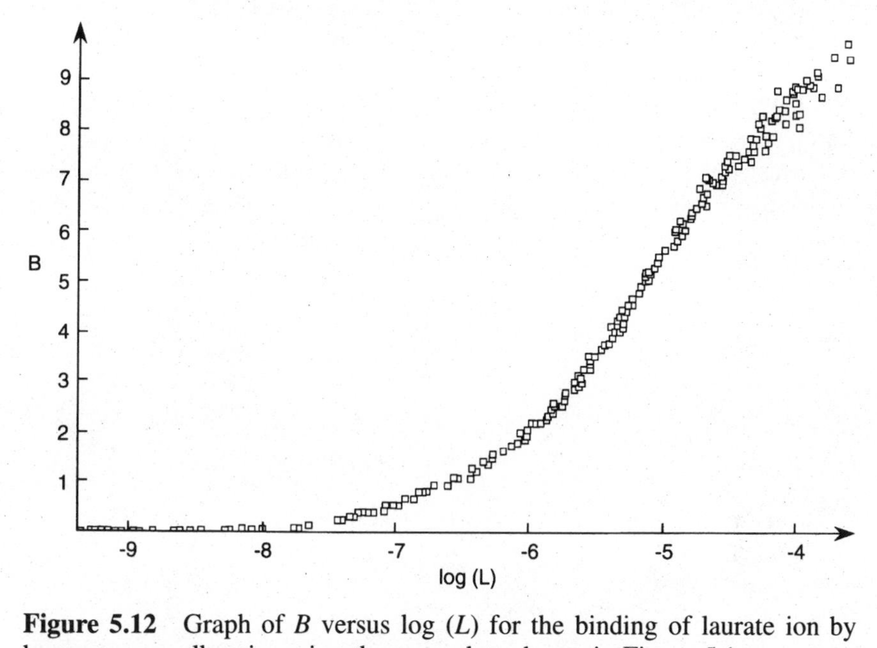

Figure 5.12 Graph of B versus log (L) for the binding of laurate ion by human serum albumin, using the same data shown in Figure 5.1.

uate either k or K from the value of (L) at half-saturation. In all other circumstances, however, there are no simple procedures for evaluating binding constants from features of the semilogarithmic plot. Nevertheless, such a graph sets the stage for the evaluation of the full range of stoichiometric binding constants by algebraic procedures. These provide the foundation for determining site binding constants for systems in which they can be established.

REFERENCES

1. A. O. Pedersen, B. Hust, S. Andersen, F. Nielsen, and R. Brodersen, *Eur. J. Biochem.*, **154,** 545–552 (1986).
2. R. Brodersen, B. Honoré, A. O. Pedersen, and I. M. Klotz, *Trends Pharm. Sci.*, **9,** 252–257 (1988).
3. E. Teng-Leary and G. B. Kohlhaw, *Biochemistry,* **12,** 2980–2986 (1973).
4. P. Skolnick, V. Moncada, J. L. Barker, and S. M. Paul, *Science,* **211,** 1448–1450 (1981).
5. I. M. Klotz, *Q. Rev. Biophys.*, **18,** 227–259 (1985).
6. *Biosoft News,* P.O. Box 10938, Ferguson, MO 63135 (Feb. 1991).

7. A. C. Jeng, T. E. Ryan, and A. E. Schamoo, *Proc. Natl. Acad. Sci. U. S. A.,* **75,** 2125–2129 (1978).
8. I. M. Klotz and D. L. Hunston, *Biochemistry,* **10,** 3065–3069 (1971).
9. J. G. Norby, P. Ottolenghi, and J. Jensen, *Anal. Biochem.,* **102,** 318–320 (1980).
10. P. Suter and J. Rosenbusch, *J. Biol. Chem.,* **251,** 5986–5991 (1976).
11. H. A. Feldman, *J. Biol. Chem.,* **258,** 12865–12867 (1983).

CHAPTER 6

NUMERICAL EVALUATIONS OF STOICHIOMETRIC BINDING CONSTANTS

With semilogarithmic graphs at hand for visualization of the full range of ligand–receptor binding data, one can then procede to fit the corresponding algebraic expression for the experimental observations,

$$B = \frac{K_1L + 2K_1K_2L^2 + \cdots + i\left(\prod_1^i K_l\right)L^i + \cdots + n\left(\prod_1^n K_l\right)L^n}{1 + K_1L + K_1K_2L^2 + \cdots + \left(\prod_1^i K_l\right)L^i + \cdots + \left(\prod_1^n K_l\right)L^n} \qquad (3.20)$$

Using appropriate algebraic fitting methods, one can determine the values of the parameters $K_1 \cdots K_i$. Let us examine the span of values encountered.

6.1. DIVALENT RECEPTORS

Some general features are manifested as soon as one moves from monovalent to divalent receptors. The magnitudes of K_i (Table 6.1) range from 10^1 to 10^8. The trend from K_1 to K_2 is most often downward (e.g., transferrins), but there are also frequent examples of an accentuation in affinity in the second stoichiometric step (e.g., cal-

TABLE 6.1. Equilibrium Constants for Ligand–Receptor Binding Manifesting Saturation

Ligand	Receptor	K_i	$\left(\frac{K_i}{K_1}\right)_{ideal}$	K_ω
Ferric ion	Ovotransferrin[a]	$K_1 = 23$	[1]	$K_\alpha = 22$
		$K_2 = 0.57$	0.25	$K_\beta = 0.6$
Zinc ion	Human transferrin[b]	$K_1 = 6.4 \times 10^7$	[1]	$K_\alpha = 6.2 \times 10^7$
		$K_2 = 0.25 \times 10^7$	0.25	$K_\beta = 0.2 \times 10^7$
Anions	Human transferrin[c]			
Bicarbonate		$K_1 = 4.6 \times 10^2$	[1]	$K_\alpha = 3.9 \times 10^2$
		$K_2 = 0.6 \times 10^2$	0.25	$K_\beta = 0.7 \times 10^2$
Phosphate		$K_1 = 1.5 \times 10^4$	[1]	$K_\alpha = 1.2 \times 10^4$
		$K_2 = 0.2 \times 10^4$	0.25	$K_\beta = 0.24 \times 10^3$
Vanadate		$K_1 = 2.8 \times 10^7$	[1]	$K_\alpha = 2.3 \times 10^7$
		$K_2 = 0.4 \times 10^7$	0.25	$K_\beta = 0.5 \times 10^7$
Leucine	Isopropylmalate synthase[d]	$K_1 = 0.48 \times 10^5$	[1]	$K_\alpha = 1.13 \times 10^5 e^{0.43\pi i}$
		$K_2 = 2.5 \times 10^5$	0.25	$K_\beta = 1.13 \times 10^5 e^{-0.43\pi i}$
Calcium ion	Calbindin[e]			
	Wild type	$K_1 = 2.2 \times 10^8$	[1]	$K_\alpha = 2.9 \times 10^8 e^{0.38\pi i}$
		$K_2 = 3.7 \times 10^8$	0.25	$K_\beta = 2.9 \times 10^8 e^{-0.38\pi i}$
	Mutant $M2$	$K_1 = 1.7 \times 10^8$	[1]	$K_\alpha = 1.7 \times 10^8$
		$K_2 = 0.005 \times 10^8$	0.25	$K_\beta = 0.01 \times 10^8$
Trifluorodihydroxypropyl phosphonate	Aspartate transcarbamoylase, half-molecule (trimer)[f]	$K_1 = 2.89 \times 10^4$	[1]	$K_\alpha = 2.89 \times 10^4$
		$K_2 = 0.0295 \times 10^4$	0.33	$K_\beta = 5.1 \times 10^2 e^{0.41\pi i}$
		$K_3 = 0.0875 \times 10^4$	0.11	$K_\gamma = 5.1 \times 10^2 e^{-0.41\pi i}$

Acetyl coenzyme A	Pyruvate carboxylase[g]	$K_1 = 2.08 \times 10^5$	[1]	$K_\alpha = 2.17 \times 10^5$
		$K_2 = 0.164 \times 10^5$	0.375	$K_\beta = 7.6 \times 10^4 e^{0.45\pi i}$
		$K_3 = 3.67 \times 10^5$	0.1667	$K_\gamma = 7.6 \times 10^4 e^{-0.45\pi i}$
		$K_4 = 0.130 \times 10^5$	0.0625	$K_\delta = 1.31 \times 10^4$
O_2	Sheep hemoglobin[h]	$K_1 = 0.1124$	[1]	$K_\alpha = 0.30e^{0.24\pi i}$
		$K_2 = 0.1974$	0.375	$K_\beta = 0.30e^{-0.24\pi i}$
		$K_3 = 0.1475$	0.1667	$K_\gamma = 0.27e^{0.70\pi i}$
		$K_4 = 1.996$	0.0625	$K_\delta = 0.27e^{-0.70\pi i}$
	Human hemoglobin[i]	$K_1 = 0.0188$	[1]	
		$K_2 = 0.0566$	0.375	
		$K_3 = 0.407$	0.1667	
		$K_4 = 4.28$	0.0625	
	Human hemoglobin cross-linked[j]	$K_1 = 0.3497$	[1]	$K_\alpha = 0.49e^{0.20\pi i}$
		$K_2 = 0.1409$	0.375	$K_\beta = 0.49e^{-0.20\pi i}$
		$K_3 = 0.2686$	0.1667	$K_\gamma = 0.39e^{0.69\pi i}$
		$K_4 = 2.802$	0.0625	$K_\delta = 0.39e^{-0.69\pi i}$
Carbamyl phosphate	Aspartate transcarbamoylase, whole molecule (hexamer)[k]	$K_1 = 1.45 \times 10^6$	[1]	$K_\alpha = 0.884 \times 10^6$
		$K_2 = 0.403 \times 10^6$	0.417	$K_\beta = 0.488 \times 10^6$
		$K_3 = 0.104 \times 10^6$	0.222	$K_\gamma = 0.0552 \times 10^6$
		$K_5 = 0.0376 \times 10^6$	0.125	$K_\delta = 0.0259 \times 10^6$
		$K_4 = 0.0151 \times 10^6$	0.067	$K_\epsilon = 0.0218 \times 10^6$
		$K_6 = 0.00821 \times 10^6$	0.028	$K_\zeta = 0.0201 \times 10^6$

[a]Ref. 1. [b]Ref. 2. [c]Ref. 3. [d]Ref. 4. [e]Ref. 5. [f]Ref. 6. [g]Ref. 7. [h]Ref. 8. [i]Ref. 9. [j]Ref. 10. [k]Ref. 11.

bindin). The anthropomorphic label *cooperative interaction* is often used to describe the latter trend. This metaphor projects human altruistic attributes on to the binding sites. Actually, at the molecular level, the changes in successive affinities are largely manifestations of conformational adaptations of the receptor macromolecule when ligand is bound and of associated alterations in interactions with solvent molecules.

For a divalent receptor, once the stoichiometric binding constants have been evaluated, one can calculate the virtual or ghost-site binding constants in a straightforward manner. Starting with equation 4.1, one can write for a divalent receptor

$$Z = 1 + K_1 L + K_1 K_2 L^2 = K_1 K_2 [(L - a_1)(L - a_2)] \tag{6.1}$$

Since the roots a_ω are related to the corresponding ghost-site binding constants (Chapter 4) by the equation

$$K_\omega = -1/a_\omega \tag{6.2}$$

equation 6.1 can be converted into

$$1 + K_1 L + K_1 K_2 L^2 = K_1 K_2 \left[\left(L + \frac{1}{K_\alpha}\right)\left(L + \frac{1}{K_\beta}\right)\right] \tag{6.3}$$

From this we can obtain

$$\frac{1}{K_1 K_2} + \frac{1}{K_2} L + L^2 = \frac{1}{K_\alpha K_\beta} + \left(\frac{1}{K_\alpha} + \frac{1}{K_\beta}\right) L + L^2 \tag{6.4}$$

A fundamental law of algebra states that “a necessary and sufficient condition that two polynomials be identical is that the corresponding coefficients be equal.” Therefore, we conclude that

$$K_1 K_2 = K_\alpha K_\beta \tag{6.5}$$

$$K_2 = \left(\frac{1}{K_\alpha} + \frac{1}{K_\beta} \right)^{-1} = \frac{K_\alpha K_\beta}{K_\alpha + K_\beta} \tag{6.6}$$

and consequently that

$$K_1 = K_\alpha + K_\beta \tag{6.7}$$

For a divalent receptor, the complementary equations giving K_α and K_β explicitly can be obtained by the following steps. Rearranging equation 6.7 we obtain

$$K_\alpha = K_1 - K_\beta \tag{6.8}$$

Using equation 6.5 to replace K_β in equation 6.8, we can rearrange the result into

$$K_\alpha^2 - K_1 K_\alpha + K_1 K_2 = 0 \tag{6.9}$$

Consequently,

$$K_\alpha = \frac{K_1 \pm \sqrt{K_1^2 - 4K_1 K_2}}{2} \tag{6.10}$$

A similar series of steps leads to

$$K_\beta = \frac{K_1 \mp \sqrt{K_1^2 - 4K_1 K_2}}{2} \tag{6.11}$$

As concrete illustrations of the relationships between stoichiometric binding constants and virtual or ghost-site binding constants, let us examine some of the results listed in Table 6.1. For example, from the experimental data for the binding of bicarbonate ion by transferrin, we find that B can be written in terms of the respective binding equations, stoichiometric or ghost site, as follows:

$$\frac{(4.6 \times 10^2)L + 2(4.6 \times 10^2)(0.6 \times 10^2)L^2}{1 + (4.6 \times 10^2)L + (4.6 \times 10^2)(0.6 \times 10^2)L^2}$$

$$= B = \frac{(3.9 \times 10^2)L}{1 + (3.9 \times 10^2)L} + \frac{(0.7 \times 10^2)L}{1 + (0.7 \times 10^2)L} \qquad (6.12)$$

Clearly, K_1 differs from K_α and K_2 differs from K_β. This feature is emphasized even more emphatically by the results for calcium binding by calbindin, for which we find the respective binding equations are

$$\frac{(2.2 \times 10^8)L + 2(2.2 \times 10^8)(3.7 \times 10^8)^2L^2}{1 + (2.2 \times 10^8)L + (2.2 \times 10^8)(3.7 \times 10^8)L^2}$$

$$= B = \frac{(2.9 \times 10^8 e^{0.38\pi i})L}{1 + (2.9 \times 10^8 e^{0.38\pi i})L} + \frac{(2.9 \times 10^8 e^{-0.38\pi i})L}{1 + (2.9 \times 10^8 e^{-0.38\pi i})L} \qquad (6.13)$$

Obviously, $K_1 = 2.2 \times 10^8$, a real number, is not the same as $K_\alpha = 2.9 \times 10^8\ e^{0.38\pi i}$, a complex, imaginary number.

Furthermore, despite the similarity in form of the right-hand sides of equations 6.12 and 6.13, the binding constants are *not* site binding constants. This fact is most obvious in calcium binding by calbindin; imaginary values for site binding constants have no meaning for real sites. The relations between K_i values, or K_ω values, and site binding constants depend on the nature of the individual sites and on the variation of their affinities with increasing binding of ligand by receptor. Specific examples will be illustrated in Chapter 10.

6.2. MULTIVALENT RECEPTORS

Stoichiometric binding constants for a range of diverse ligand–receptor complexes are assembled in Tables 6.1 and 6.2. For the enzymes and the hemoglobins listed, the saturation values for ligand bound at the active site are known from structural information. In contrast, for the serum albumins, semilogarithmic graphs show clearly that saturation of receptor by ligand has not been attained (see Fig. 5.12). Nevertheless, values of K_i can be obtained from binding data up to the highest experimental results for moles of bound ligand B.

In most of these examples, the observed stoichiometric binding con-

TABLE 6.2 Equilibrium Constants for Ligand–Receptor Binding Not Attaining Saturation

Ligand	Receptor	K_i	$\left(\frac{K_i}{K_{i-1}}\right)_{\text{ideal}}$	K_ω
Hexanoate	Human	$K_1 = 7.08 \times 10^4$	—	$K_\alpha = 3.59 \times 10^4$
ion[a]	serum	$K_2 = 1.81 \times 10^4$	0.5	$K_\beta = 3.40 \times 10^4$
	albumin	$K_3 = 8.03 \times 10^2$	0.66	$K_\gamma = 9.1 \times 10^2$
		$K_4 = 4.20 \times 10^1$	0.75	$K_\delta = 3.0 \times 10^2\, e^{0.46\pi i}$
		$K_5 = 2.32 \times 10^3$		$K_\epsilon = 3.0 \times 10^2\, e^{-0.46\pi i}$
Octanoate	Human	$K_1 = 1.62 \times 10^6$	—	$K_\alpha = 1.58 \times 10^6$
ion[b]	serum	$K_2 = 3.49 \times 10^4$	0.5	$K_\beta = 2.93 \times 10^4$
	albumin	$K_3 = 6.10 \times 10^3$	0.65	$K_\gamma = 1.10 \times 10^3$
		$K_4 = 4.67 \times 10^3$	0.74	
		$K_5 = 9.87 \times 10^2$		
		$K_6 = 1.38 \times 10^4$		
		$K_7 = 2.83 \times 10^0$		
		$K_8 = 3.10 \times 10^5$		
		$K_8 = 2.60 \times 10^3$		
Laurate	Human	$K_1 = 8.31 \times 10^6$	—	
ion[b]	serum	$K_2 = 1.67 \times 10^6$	0.5	
	albumin	$K_3 = 2.88 \times 10^5$	0.65	
		$K_4 = 6.18 \times 10^5$	0.74	
		$K_5 = 1.07 \times 10^2$		
		$K_6 = 1.98 \times 10^8$		
		$K_7 = 9.25 \times 10^1$		
		$K_8 = 1.43 \times 10^7$		
		$K_9 = 1.86 \times 10^3$		
		$K_{10} = 1.94 \times 10^4$		
Diflunisal[c]	Human	$K_1 = 5 \times 10^5$	—	
	serum	$K_2 = 2.2 \times 10^5$	0.5	
	albumin	$K_3 = 1.2 \times 10^5$	0.6	
		$K_4 = 7.5 \times 10^4$	0.7	
		$K_5 = 4.7 \times 10^4$		
		$K_6 = 2.8 \times 10^4$		
		$K_7 = 1.6 \times 10^4$		
		$K_8 = 7.7 \times 10^3$		
		$K_9 = 4.4 \times 10^3$		
		$K_{10} = 2.7 \times 10^3$		
		$K_{11} = 1.8 \times 10^3$		
		$K_{12} = 1.2 \times 10^3$		
		$K_{13} = 7.3 \times 10^2$		
		$K_{14} = 4.2 \times 10^2$		
		$K_{15} = 1.8 \times 10^2$		

TABLE 6.2. (*Continued*)

Ligand	Receptor	K_i	$\left(\frac{K_i}{K_{i-1}}\right)_{\text{ideal}}$	K_ω
Azosulfa-thiazole[d]	Bovine serum albumin	$K_1 = 1.25 \times 10^5$	—	—
		$K_2 = 4.50 \times 10^4$	0.5	
		$K_3 = 2.16 \times 10^4$	0.6	
		$K_4 = 1.17 \times 10^4$	0.7	
		$K_5 = 6.70 \times 10^3$		
		$K_6 = 3.99 \times 10^3$		
		$K_7 = 2.43 \times 10^3$		
		$K_8 = 1.51 \times 10^3$		
		$K_9 = 9.5 \times 10^2$		
		$K_{10} = 6.0 \times 10^2$		
		$K_{11} = 3.8 \times 10^2$		
		$K_{12} = 2.4 \times 10^2$		
		$K_{13} = 1.54 \times 10^2$		
		$K_{14} = 7.0 \times 10^1$		
		$K_{15} = 4.6 \times 10^1$		
		$K_{16} = 2.5 \times 10^1$		

[a]Ref. 12. [b]Ref. 13. [c]Ref. 14. [d]Ref. 15.

stants decrease with an increase in bound ligand. In contrast, for the hemoglobins, successive K_i values increase. Still different behavior is found with pyruvate carboxylase and with the serum albumins, where a general downward trend is interrupted by a sudden accentuation in affinity for ligand.

To obtain insights into the significance of trends in K_i values, we need to have some reference standard with which to make a comparison. In physicochemical studies of gases, one uses the behavior of an ideal gas as a standard and interprets the behavior of real gases by examining deviations from ideal behavior. A similar procedure is advantageous for ligand binding. Let us find, therefore, how K_i varies with i for ideal binding, i.e., when all the sites are identical and invariant in affinity. For this objective, it is convenient again to return to the exposition of the origin of virtual or ghost-site binding constants (Chapter 4).

For the special case of a polynomial for which all ω roots are identical, equation 4.1 reduces to

$$Z = \left(\prod K_l\right)(L - a)^n \tag{6.14}$$

Inserting this relation into equations 4.5 and 4.6, and defining

$$(-1/a) = \kappa \tag{6.15}$$

we obtain a new form of equation 4.7 for B

$$B = \frac{n\kappa L}{1 + \kappa L} \tag{6.16}$$

If we compare this relationship with equation 2.14, we see that

$$\kappa = k = K \tag{6.17}$$

In other words, when binding is ideal, the ghost binding constant is identical with the site binding constant and with the stoichiometric constant for an isolated monovalent receptor.

Nevertheless, even in ideal binding behavior, with all sites having identical, invariant binding affinities, it is not true that each successive stoichiometric constant is equal to k (or K),

$$K_1 \neq k \neq K_2 \cdots \neq K_i \tag{6.18}$$

There are, however, explicit relationships between K_i and k, which we are now in a position to derive.

Let us consider first a divalent receptor. For ideal binding we can write a more explicit form of equation 6.1:

$$Z = 1 + K_1 L + K_1 K_2 L^2 = K_1 K_2 \left(L + \frac{1}{\kappa}\right)^2 \tag{6.19}$$

The right-hand side of equation 6.19 can be expanded so that we may write

$$\frac{1 + K_1 L + K_1 K_2 L^2}{K_1 K_2} = \frac{\kappa^2 L^2 + 2\kappa L + 1}{\kappa^2} \tag{6.20}$$

From this it follows that

$$\frac{1}{K_1K_2} + \frac{1}{K_2} L + L^2 = \frac{1}{\kappa^2} + \frac{2}{\kappa} L + L^2 \tag{6.21}$$

We now have two different polynomial expressions for the same Z. Recalling that in such a case a necessary and sufficient condition that two polynomials be identical is that the corresponding coefficients be equal, we conclude that

$$\frac{1}{K_2K_2} = \frac{1}{\kappa^2} \tag{6.22}$$

$$\frac{1}{K_2} = \frac{2}{\kappa} \tag{6.23}$$

Taking cognizance of the fact that $\kappa = k$, we restate equations 6.22 and 6.23 as

$$(K_2)_{\text{ideal}} = \tfrac{1}{2}k \tag{6.24}$$

$$(K_1K_2)_{\text{ideal}} = k^2 \tag{6.25}$$

and then add

$$(K_1)_{\text{ideal}} = 2k \tag{6.26}$$

Thus we have explicit relations between the stoichiometric binding constants and the site binding constants for a divalent receptor with identical, invariant site affinities for ligand.

It is probably apparent that this type of analysis is not limited to ideal divalent receptors. For the general case of an n-valent receptor, in place of equation 6.19 we obtain

$$1 + K_1L + K_1K_2L^2 + \cdots + (K_1K_2\cdots K_i)L^i + \cdots + (K_1K_2\cdots K_n)L^n$$
$$= (K_1K_2\cdots K_n)\left(L + \frac{1}{\kappa}\right)^n \tag{6.27}$$

This leads to the following general relationships between stoichiometric and site binding constants for an n-valent receptor with identical invariant site affinities for ligand (see Appendix A3):

$$(K_1)_{\text{ideal}} = \frac{n}{1}\, k$$

$$(K_2)_{\text{ideal}} = \frac{n-1}{2}\, k$$

$$(K_3)_{\text{ideal}} = \frac{n-2}{3}\, k$$

$$\vdots$$

$$(K_i)_{\text{ideal}} = \frac{n-(i-1)}{i}\, k \tag{6.28}$$

$$\vdots$$

$$(K_n)_{\text{ideal}} = \frac{1}{n}\, k$$

With equation 6.28, we can write explicit expressions in any specific case for the successive stoichiometric binding constants $K_1 \cdots K_n$ for a ligand–receptor complex with sites of identical, invariant affinities. The actual variation of K_i with i can then be compared with that of the ideal state. Representative trends will be presented in Chapter 7.

6.3. SIMPLE GRAPHICAL EVALUATION OF K_1

For comparison of affinities of a receptor for different ligands or of affinities of different receptors for a common ligand, one can use the respective first stoichiometric binding constant K_1. To obtain K_1, one does not need to carry out a complete statistical fit to equation 3.20 to determine the best values of all of the constants. It is almost always possible to evaluate K_1 by a simple graphical procedure (16).

As is evident from equation 3.20, after we factor out (L) from the numerator and transfer it to the left-hand side of the equation, we can take the limit of this equation and obtain

$$\lim_{(L)\to 0} \frac{B}{(L)} = K_1 \tag{6.29}$$

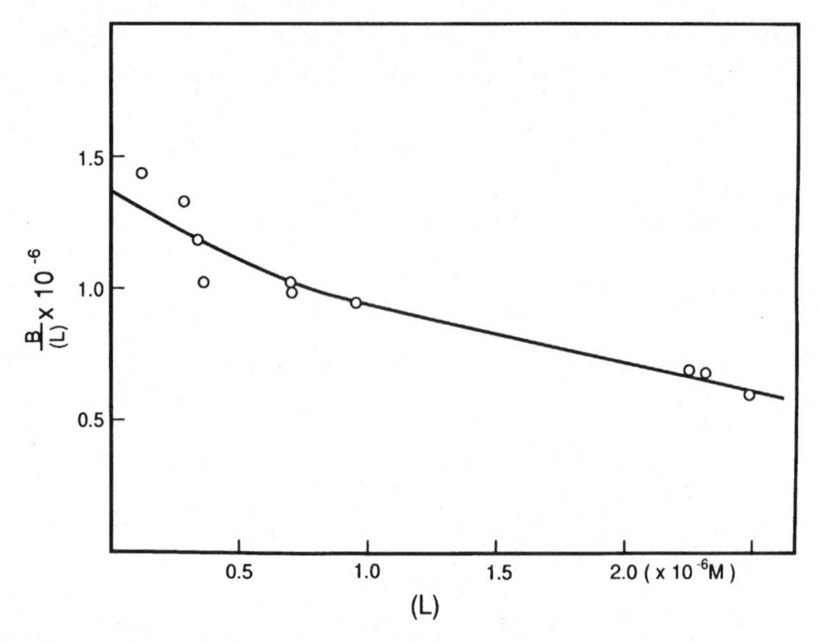

Figure 6.1 Evaluation of the first stoichiometric binding constant for the binding of carbamoyl phosphate by aspartate transcarbamoylase by extrapolation of the values of $B/(L)$ to the limiting value as free ligand concentration approaches zero. Data from Ref. 16.

Usually, it is possible to measure binding with good precision at low concentrations of ligand. Hence an extrapolation, such as that illustrated for a specific example in Figure 6.1, provides a value for the first stoichiometric binding constant.

REFERENCES

1. J. W. Donovan, R. A. Beardslee, and K. D. Ross, *Biochem J.,* **153,** 631–639 (1976).
2. W. R. Harris, *Biochemistry,* **22,** 3920–3926 (1983).
3. W. R. Harris, *Biochemistry,* **24,** 7412–7418 (1985).
4. E. Teng-Leary and G. B. Kohlhaw, *Biochemistry,* **12,** 2980–2986 (1973).
5. S. Linse, P. Broden, T. Drakenberg, E. Thulin, P. Sellers, K. Elmden, T. Grundstrom, and S. Forsén, *Biochemistry,* **26,** 6723–6735 (1987).
6. J. A. Ridge, M. F. Roberts, M. H. Schaffer, and G. R. Stark, *J. Biol. Chem.,* **251,** 5966–5975 (1976).

7. W. H. Frey, and M. F. Utter, *J. Biol. Chem.,* **252,** 51–56 (1977).
8. F. J. W. Roughton, A. B. Otis, and R. L. J. Lyster, *Proc. R. Soc. London B,* **144,** 29–54 (1955).
9. K. Imai, *Adv. Enzymol.,* **232,** 559–576 (1994).
10. S. Miura, M. Ikeda-Saito, T. Yonetani, and C. Ho, *Biochemistry,* **26,** 2149–2155 (1987).
11. I. M. Klotz and D. L. Hunston, *Proc. Natl. Acad. Sci. USA,* **74,** 4959–4963 (1977).
12. B. Honoré and R. Brodersen, *Anal. Biochem.,* **171,** 55–66 (1988).
13. A. O. Pedersen, B. Hust, S. Andersen, F. Nielsen, and R. Brodersen, *Eur. J. Biochem.,* **154,** 545–552 (1986).
14. B. Honoré and R. Brodersen, *Mol. Pharmacol.,* **25,** 137–150 (1984).
15. I. M. Klotz, F. M. Walker, and R. B. Pivan, *J. Am. Chem. Soc.,* **68,** 1486–1490 (1946).
16. I. M. Klotz in T. E. Creighton, ed., *Protein Function: A Practical Approach,* IRL Press, Oxford, UK, 1989, Chap. 2.

CHAPTER 7

AFFINITY PROFILES

For numerical comparisons of the trends in actual K_i values with those for corresponding ideal receptors, ratios of $(K_i/K_1)_{\text{ideal}}$ have been listed in Table 6.1. Thus one can see that for ferric ion–ovotransferrin, for example, the second stoichiometric constant K_2 is decreased markedly, more than would be expected for an ideal two-site receptor. This does *not* mean that the actual receptor contains two sites of different, invariant affinities, with site constants k_1 and k_2 and $k_2 \ll k_1$. Rather, as will be shown in Chapter 10, this receptor could contain initially two identical sites with identical initial site binding constants; but if the affinity of each site changed when the other site was occupied by ligand, then the binding isotherm would still fit the equation

$$B = \frac{22L}{1 + 22L} + \frac{0.6L}{1 + 0.6L} \tag{7.1}$$

In equation 7.1, 22 and 0.6 are K_α and K_β, respectively, and are not assignable to any individual site.

For graphical presentation of trends in actual K_i values, it is convenient to have at hand a rearranged form of equation 6.28 for ideal receptors:

$$i(K_i)_{\text{ideal}} = (n + 1)k - ki \tag{7.2}$$

This normalized equation tells us that for an ideal receptor–ligand complex a graph of $i(K_i)_{\text{ideal}}$ versus i, an affinity profile, will present a straight line, a convenient reference form (1).

An example of actual behavior of a divalent receptor is illustrated by leucine–isopropylmalate synthase (2) presented in Figure 7.1*A*. It is immediately evident from an affinity profile that after 1 mole of ligand has been bound, the affinity in the second sequential step is markedly accentuated. As will be seen in Chapter 10, this can be rationalized in molecular terms if the binding of ligand at either site increases the affinity for ligand at the remaining unoccupied site.

A complementary example is illustrated in Figure 7.1*B*, an affinity profile for the binding of ferric ion by ovotransferrin (3). Here, the normalized actual affinity iK_i drops precipitously in the second stoichiometric step, compared to the value expected for $i(K_i)_{\text{ideal}}$. This trend could be a manifestation of the existence of two binding sites

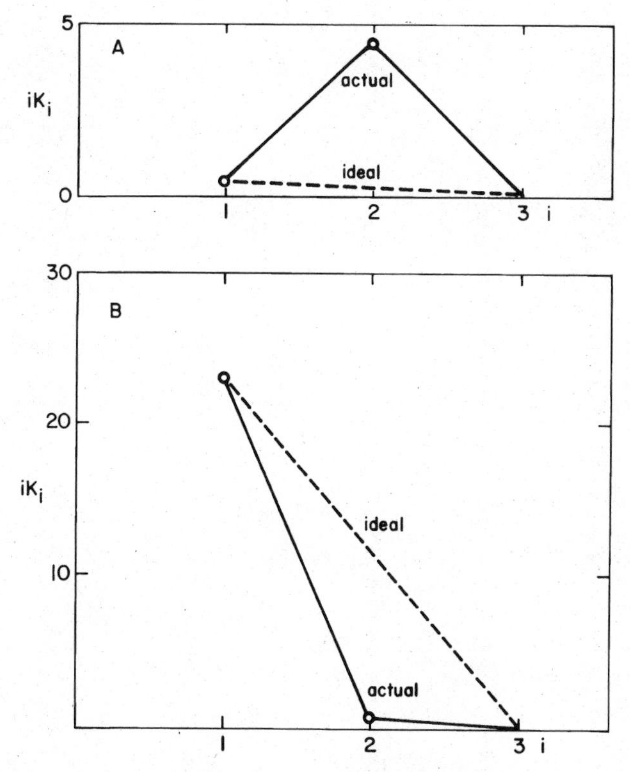

Figure 7.1 Affinity profiles for the binding of leucine by isopropylmalate synthase (*A*), and for binding of iron by ovotransferrin (*B*).

with different initial affinities that are invariant with ligand occupancy. On the other hand, the same trend would be seen if the two empty sites have initially identical affinities, which decrease markedly when the companion site is occupied by ligand. From thermodynamic binding measurements alone, one cannot discriminate between these two possibilities. With additional probes, it may be possible to do so. For iron–ovotransferrin the individual site binding constants have been resolved (3) and are $k_1 = 13$, $k_2 = 10$, $k_{1,2} = 1.0$ and $k_{2,1} = 1.3$.

Ovotransferrin binding of iron falls into a class often called *half-of-sites* reactivity, i.e., the first iron is taken up by either of the two almost equivalent unoccupied sites and then the affinity of the residual site is markedly attenuated, regardless of which one it was in the totally unoccupied protein.

From the stoichiometric constants in Table 6.1 for the binding of trihydroxypropyl phosphonate (an analogue of carbamyl phosphate) by the half-molecule [$(CR)_3$] form of aspartate transcarbamoylase (4) (normally in the $(CR)_6$ state), one can prepare the graph shown in Figure 7.2. In this case one can see strikingly that the initial large

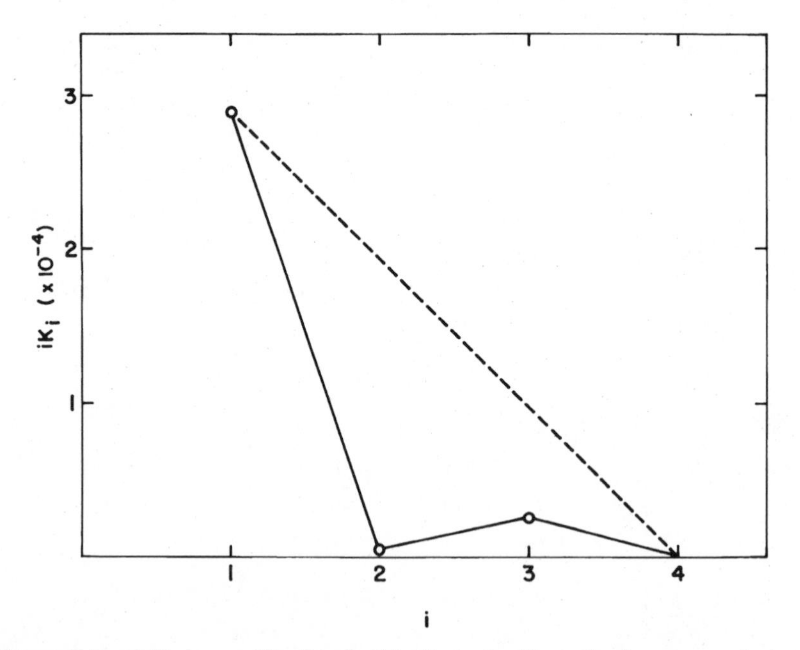

Figure 7.2 Affinity profile for the binding of trifluorohydroxypropyl phosphonate by aspartate transcarbamoylase (*solid line*) compared with an ideal profile (*broken line*).

negative slope from $i = 1$ to $i = 2$ is followed by a sharp change to a positive slope from $i = 2$ to $i = 3$. Thus the nature of the receptor and its interactions with ligand undergo a profound change during the third stoichiometric binding step. This kind of resolution is not evident in a Scatchard plot. The curves seen in such a single-reciprocal graph, convex or concave to the coordinate axes, are ascribed broadly to cooperative or antagonistic interactions but do not reveal the changes that occur at successive steps in binding of ligand.

An even more striking variable behavior is illustrated in Figure 7.3, the affinity profile for the tetravalent receptor pyruvate carboxy-

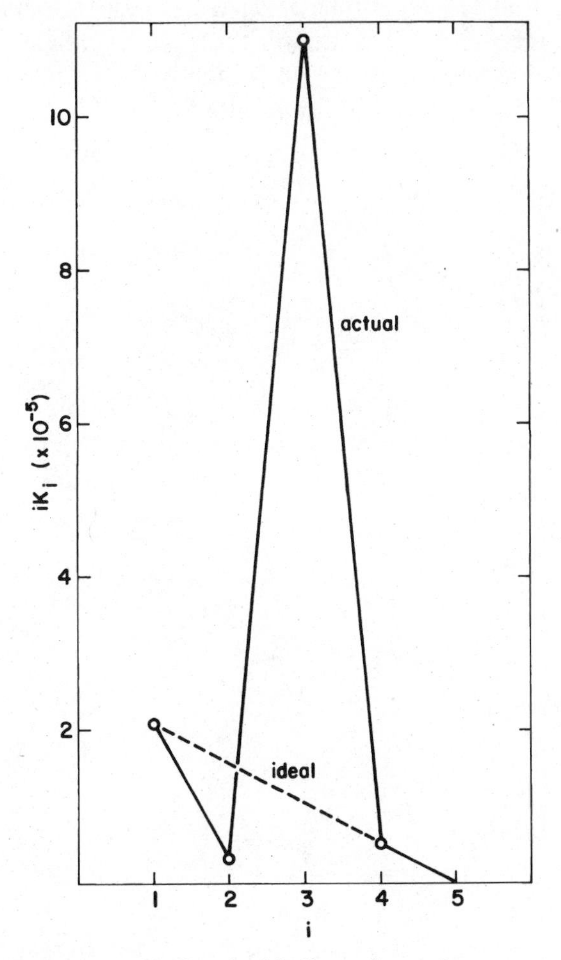

Figure 7.3 Affinity profile for the binding of acetyl coenzyme A by pyruvate carboxylase.

lase binding acetyl coenzyme A (5). Pyruvate carboxylase displays a markedly accentuated affinity for acetyl coenzyme A in the third stoichiometric step of the uptake of ligand. That is followed by a steep drop in affinity in the fourth stoichiometric step, so that the normalized affinity $4K_4$ falls back to the initial line for an ideal trend.

Another tetravalent system, of wide interest, is hemoglobin (6) whose affinity profile is illustrated in Figure 7.4. The accentuations in affinity at each step are so strong that the comparison line for ideal behavior is essentially invisible because it lies essentially along the x-axis.

When the saturation value for moles of bound ligand n cannot be established from the experimental data, equation 7.2 cannot be used to draw an ideal line in an affinity profile. Nevertheless, the terminal point on the x-axis must be above the highest value observed for B. Figure 7.5 presents an affinity profile for binding of hexanoate by

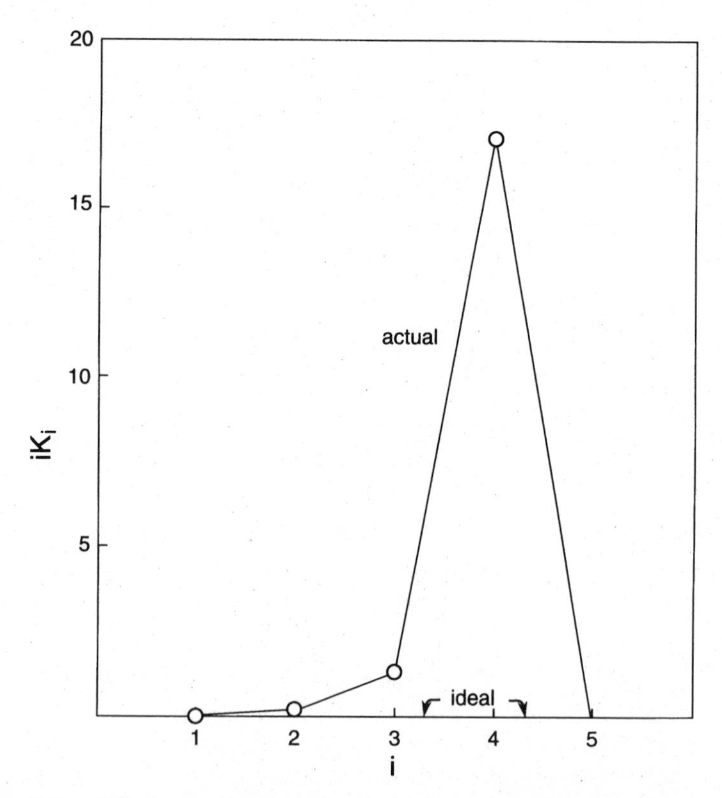

Figure 7.4 Affinity profile for uptake of O_2 by hemoglobin. On the scale displayed, the ideal line cannot be distinguished from the abscissa axis.

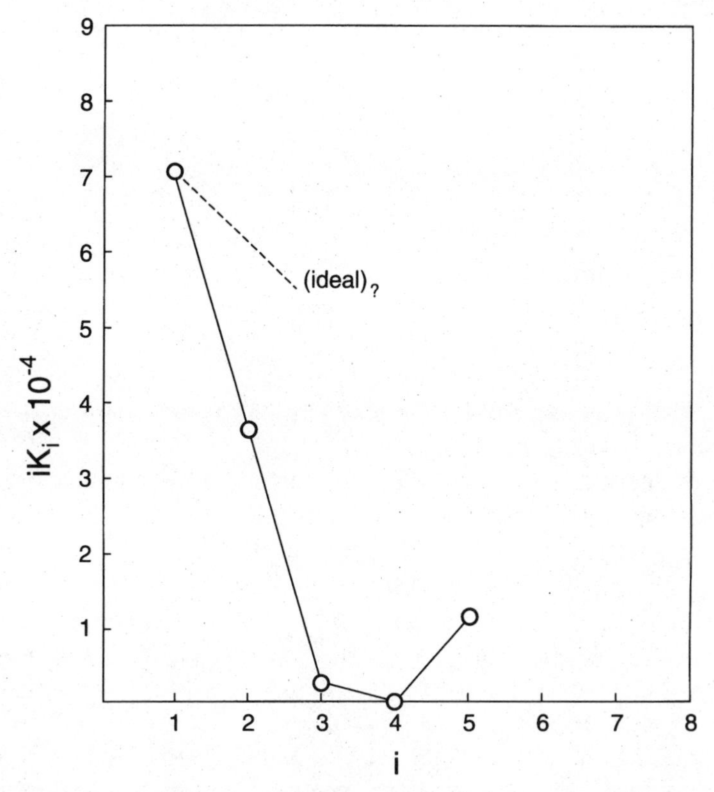

Figure 7.5 Affinity profile for binding of hexanoate ions by human serum albumin. Since n, the saturation value for this ligand–receptor complex, has not been established, the line for $i(K_i)_{\text{ideal}}$ must be to the right of the one shown.

human serum albumin. Since the value of n, the number of moles of ligand bound by the receptor at saturation has not been determined experimentally, it is not possible to draw a line for $i(K_i)_{\text{ideal}}$. The broken line shown is a minimal guess based on the conjecture that $n > 6$. Regardless of the position of the actual line for $i(K_i)_{\text{ideal}}$, it is evident from the slopes of the connecting lines that a marked accentuation above the ideal occurs between K_4 and K_5, and a smaller one probably appears in the step of K_3 to K_4.

Although equation 7.2 cannot be used to draw an ideal line when saturation is indeterminate, it is still possible to obtain an approximate quantitative measure of deviations from ideal behavior in the successive stoichiometric steps. From equation 7.2, it follows that

$$\left(\frac{K_i}{K_{i-1}}\right)_{\text{ideal}} = \left(\frac{n-i+1}{n-i+2}\right)\left(\frac{i-1}{i}\right) \tag{7.3}$$

During the initial stoichiometric steps, $i < n$, hence

$$\left(\frac{K_i}{K_{i-1}}\right)_{\text{ideal}} \simeq (1)\,\frac{i-1}{i} \tag{7.4}$$

This relation allows one to compute entries for the first few stoichiometric steps (see Table 6.2).

From these reference values, we can see that with each of the anions bound by human serum albumin the actual values of K_2/K_1 are all substantially smaller than the ideal ratio of 0.5. Values for $(K_i/K_{i-1})_{\text{ideal}}$ increase progressively as i becomes 3, 4, etc., and reach a maximum as $i \longrightarrow n/2$. If n is a very large number, then the maximum value of $(K_i/K_{i-1})_{\text{ideal}}$ approaches 1. In all cases, the ideal ratio drops downward as i values progress beyond $n/2$ toward n, and approaches 0.5 at $i = n$. Thus the successive actual values of K_i/K_{i-1} (see Table 6.2) can be compared with the expected ideal ratios even when n, the moles of ligand that saturate the receptor, cannot be specified from the binding measurements.

Stoichiometric binding constants reflect the nature of the interactions between ligand and receptor at different stages of occupancy. Each entity RL_i is a different species from its neighboring cohort; hence it is not surprising that K_i should differ from K_{i-1} by an amount less than or greater than the statistics of identical sites. One would expect that formation of a complex between L and RL_i would be accompanied by mutual accommodations in structure of each participant to its new partner and by adaptations in interactions with solvent molecules. Furthermore, there is no intrinsic reason why such multiple interactions should be the same in successive stoichiometric steps.

One can construct molecular scenarios specifying the behavior of sites on the receptor that can lead to the observed experimental values of the stoichiometric binding constants. Representative examples will be elaborated in Chapter 10. Thermodynamic constructs provide a sieve through which any postulated molecular model must be able to pass. A choice between alternative scenarios, however, requires additional insights obtainable from probes of molecular structure and dynamics.

REFERENCES

1. I. M. Klotz and D. L. Hunston, *Arch. Biochem. Biophys.,* **193,** 314–328 (1979).
2. E. Teng-Leary and G. R. Kohlhaw, *Biochemistry,* **12,** 2980–2986 (1973).
3. J. W. Donovan, R. A. Beardslee, and K. O. Ross, *Biochem. J.,* **153,** 631–639 (1976).
4. J. A. Ridge, M. F. Roberts, M. H. Schaffer, and G. R. Stark, *J. Biol. Chem.,* **251,** 5966–5975 (1976).
5. W. H. Frey and M. F. Utter, *J. Biol. Chem.,* **252,** 51–56 (1977).
6. K. Imai, *Methods Enzymol.,* **232,** 559–655 (1994).

CHAPTER 8

THERMODYNAMIC PERSPECTIVES

Since the strength of binding of ligand to receptor is expressed in the equilibrium constant, once that parameter has been measured one has an entry into energetic quantities through the equation

$$\Delta G^\circ = -RT \ln Q_{\text{equilibrium}} \tag{8.1}$$

where Q represents the quotient of equilibrium concentrations K_i, k_j or K_ω; and ΔG° is the standard free energy for the chemical transformation associated with the corresponding equilibrium constant. The stoichiometric equilibrium constant K_i is the most appropriate one for a clear and general visualization of the states to which ΔG° refers.

8.1. VARIATION OF ΔG° IN A SERIES OF RELATED LIGANDS

For making these comparisons the focus will be on ΔG_1°, which corresponds to the first stoichiometric equilibrium constant K_1. As pointed out in Chapter 6, it is almost always possible to determine K_1 from a simple graphical extrapolation of binding data at low ligand concentrations. A comparison of ΔG_1° values quantitatively appraises the affinity of a particular receptor for the first ligand of each structure in the series that becomes linked to it.

From the data in Table 6.1 it is possible to calculate the free energy

TABLE 8.1 Free Energy Changes for Ligand–Receptor Binding Manifesting Saturation[a]

Ligand	Receptor	ΔG_i° (cal mol^{-1})	$\Delta G_i^\circ - \Delta G_{i-1}^\circ$ (cal mol^{-1})
Ferric ion	Ovotransferrin	−1,830	—
		330	2,160
Zinc ion	Human transferrin	−10,600	—
		−8,700	1,900
Anions	Human transferrin		
Bicarbonate		−3,630	—
		−2,430	1,200
Phosphate		−5,700	—
		−4,510	1,190
Vandate		−10,170	—
		−9,010	1,160
Leucine	Isopropylmalate synthase	−6,390	—
		−7,370	−980
Calcium ion	Calbindin		
	Wild type	−11,390	—
		−11,700	−310
	Mutant $M2$	−11,230	—
		−7,780	3,450
Trifluorodihydroxy-propyl phos-phonate	Aspartate transcar-bamoylase half-molecule (trimer)	−6,090	—
		−3,370	2,720
		−4,020	−650
Acetyl coenzyme A	Pyruvate carboxylase	−7,260	—
		−5,750	1,510
		−7,600	−1,850
		−5,620	1,980
O_2	Human hemoglobin	2,360	—
		1,700	−660
		530	−1,170
		−860	−1,390
Carbamyl phosphate	Aspartate transcar-bamoylase, whole molecule (hexa-mer)	−8,410	—
		−7,650	760
		−6,850	800
		−6,240	610
		−5,700	540
		−5,340	360

[a]For corresponding stoichiometric equilibrium constants see Table 6.1.

changes ΔG_1° for the binding of various anions by Fe transferrin (Table 8.1). For the first mole of bound ligand the values are:

Ligand	ΔG_1° (cal mol^{-1})
Bicarbonate	−3,630
Phosphate	−5,700
Vanadate	−10,170

In this series, the observed trend correlates well with the affinity of Fe(III) for these anions in nonprotein solutions and hence confirms the presumption that in transferrin these anions are bound to the iron centers.

Small inorganic ions are bound also by proteins that have no metal centers. Thus serum albumin forms complexes with a wide range of ions with different affinities. For example, comparing chloride and thiocyanate ions, one finds the ΔG_1° values to be (Table 8.2):

TABLE 8.2 Free Energies, Enthalpies, and Entropies in Formation of Ligand–Receptor Complexes

Ligand	Receptor	ΔG° (cal mol^{-1})	ΔH° (cal mol^{-1})	ΔS° (cal mol^{-1} deg^{-1})
Chloride	Serum albumin[a]	−2,200	400	8
Thiocyanate		−4,100	0	14
Cupric ion	Serum albumin[b]	−5,900	2,800	29
	Serum albumin[c]	−6,500		
	β-lactoglobulin[c]	−5,800		
	Lysozyme[c]	−4,400		
Acetate	Bovine serum albumin[d]	−3,500	0	9
Valerate		−4,500	0	12
Caproate		−4,600	−820	10
Heptanoate		−5,300	−3,300	4
Caprylate		−6,100	−4,700	1
Caproate	Human serum albumin[e]	−6,800		
Caprylate		−8,700		
Methyl orange	Serum albumin[f]	−6,400	−2,100	14
Azosulfathiazole	Serum albumin[f]	−7,150	−2,000	17
Salicylate	Serum albumin[g]	−6,650		
Diflunisal		−8,050		
Sulfamethizole	Serum albumin[h]	−6,000		
Bilirubin	Serum albumin[i]	−11,200		

TABLE 8.2 (*Continued*)

Ligand	Receptor	ΔG° (cal mol^{-1})	ΔH° (cal mol^{-1})	ΔS° (cal mol^{-1} deg^{-1})
Ceftriaxone	Serum albumin[j]	−6,500		
Hexadecylphosphorylcholine	Phospholipase A_2[k]	−6,800	−360	24
Cytidine monophosphate	Ribonuclease A[l]	−7,500	−14,100	−22
Cytidine 2′-monophosphate	Ribonuclease A[m]	−6,700	−10,600	−13
Cytidine triphosphate	Aspartate transcarbamoylase[n]	−6,900	−13,500	−22
Diphosphopyridine nucleotide	Glyceraldehyde phosphate dehydrogenase[o]	−10,000		
FK 506	FK binding protein[p]	−12,300	−17,200	−16
Dinitrophenyl-lysine	Anti-DNP antibody[q]	−9,700	−13,800	−14
Tripeptide	Vancomycin[l]	−7,200	−11,500	−14
Propranolol	Human α-acid glycoprotein[r]	−7,700		
Propafenone, enantiomer	Human α-acid glycoprotein[s]			
R–		−8,081		
S–		−8,122		
Anions	Sulfate-binding protein[t]			
SO_4^{2-}			−9,300	
SeO_4^{2-}			−7,100	
CrO_4^{2-}			−8,700	
$HCrO_4^-$			−7,400	
HPO_4^{2-}			−1,600	
$H_2PO_4^-$			~0	
Methylmannoside	Concanavalin A[l]	−5,000	−7,100	−7
Dimannoside		−5,300	−7,000	−6
Trimannoside		−7,500	−10,700	−11
Trimannoside	*Dioclea lectin*[l]	−8,200	−13,000	−16
O-antigenic oligosaccharides	Monoclonal antibody Se155-4[u]			
3-mer		−7,300	−4,900	8.0
8-mer		−7,800	−8,200	−1.3
12-mer		−7,700	−10,600	−9.5
16-mer		−8,300	−10,900	−8.7
20-mer		−7,800	−17,000	−32
Lysozyme	Antibody Fv fragment[v]			
	Wild type	−11,400	−20,300	−30
	Mutant-1	−8,500	−18,300	−33
	Mutant-2	−8,800	−13,700	−16
	Mutant-11	−8,600	−21,800	−44
	Mutant-17	−8,700	−20,700	−40

TABLE 8.2 (*Continued*)

Ligand	Receptor	ΔG° (cal mol^{-1})	ΔH° (cal mol^{-1})	ΔS° (cal mol^{-1} deg^{-1})
Human IgG	Human IgM monoclonal rheumatoid factor[w]	<(−13,000)	−13,000	>0
λ repressor	Operator O_R[x]			
	Site 1	−11,600		
	Site 2	−10,100		
	Site 3	−10,100		
Cro protein	DNA[y]			
Wild type	operator OR1	−15,400	1,200	58
	operator OR3	−16,100	800	59
	nonoperator	−13,500	3,000	57
	nonspecific	−9,700	4,400	49
	OR3 mutant · 2TA	−12,300	7,900	70
	OR3 mutant · 2UA	−10,300	5,700	55
	OR3 mutant · 4AT	−12,200	5,800	63
	OR3 mutant · 2AU	−16,000	1,200	60
Mutant	OR3	−11,700	5,700	60
Val 27	OR3 mutant · 2TA	−14,000	8,900	79
	OR3 mutant · 2UA	−12,200	7,300	68
	OR3 mutant · 2AU	−11,600	4,700	57
Human growth hormone (hGH)	hGH receptor[z]	−12,300		

[a]Ref. 1. [b]Ref. 2. [c]Ref. 3. [d]Ref. 4. [e]Ref. 5. [f]Ref. 6 and 7. [g]Ref. 8. [h]Ref. 9. [i]Ref. 10. [j]Ref. 11. [k]Ref. 12. [l]Ref. 13. [m]Ref. 14. [n]Ref. 15. [o]Ref. 16. [p]Ref. 17. [q]Ref. 18. [r]Ref. 19. [s]Ref. 20. [t]Ref. 21. [u]Ref. 22. [v]Ref. 23. [w]Ref. 24. [x]Ref. 25 and 26. [y]Ref. 27. [z]Ref. 28.

Ligand	ΔG_1° (cal mol^{-1})
Cl^-	−2200
SCN^-	−4100

The substantially larger interaction free energy of thiocyanate ion is probably a reflection of its more polarizable character. There is strong evidence that anions are bound near cationic site chains of serum albumin (29).

It should be pointed out that it is not meaningful to say that the interaction energy of albumin with thiocyanate compared with chloride is increased by 87% ((4100 – 2200/2200) × 100). Such a claim implies that a ΔG° of zero corresponds to no interaction between ligand and receptor. That presumption is clearly wrong, for a ΔG° of zero corresponds to an equilibrium constant of 1 (see equation 8.1). Depending on the concentrations of ligand and receptor, if $K_1 \simeq 1$, a substantial fraction of receptor may be in the liganded state RL_1.

Among organic anions, the closest structural relatives are stereoisomers. It has long been known (30) that in interactions with specific receptors (e.g., specific antibodies or enzymes) marked selectivity occurs. Similarly, selectivity is also evident in binding by nonspecific receptors. Quantitative studies of the binding by serum albumin of the optically isomeric (D and L) forms of the dye-labeled amino acid (31),

$$(CH_3)_2N-C_6H_4-N{=}N-C_6H_4-\overset{O}{\overset{\|}{C}}-NH-\overset{H}{\overset{|}{C}}(C_6H_5)-CO_2^-$$

give ΔG_1° values of -7630 and -7200 cal mol^{-1} for the L and D isomers, respectively. Obviously, these optical isomers do not face the receptor surface with exactly the same disposition of groups. The value -430 cal mol^{-1} for $\Delta G_1^\circ(\text{L}) - \Delta G_1^\circ(\text{D})$ is an increment in affinity comparable to that seen with an added CH_2 group in some base structures.

The classic studies of Landsteiner (32) demonstrate (qualitatively) the extraordinary ability of specific antibodies to discriminate between small aromatic molecules with minor differences in structure, e.g.,

$$-N{=}N-C_6H_4-CO_2^-, \quad -N{=}N-C_6H_4-SO_3^-,$$

$$-N{=}N-C_6H_4-AsO_3H^-$$

Quantitative experiments have generated values for the free energies of the respective interactions (33). With an antibody homologous to the azophenylarsonate group, these studies show a strong affinity for the anionic dye,

$$HO-C_6H_4-N{=}N-C_6H_4-AsO_3H^-$$

for which $\Delta G^\circ = -7700$ cal mol^{-1}; the same γ-globulin does not significantly bind the corresponding sulfonate anion

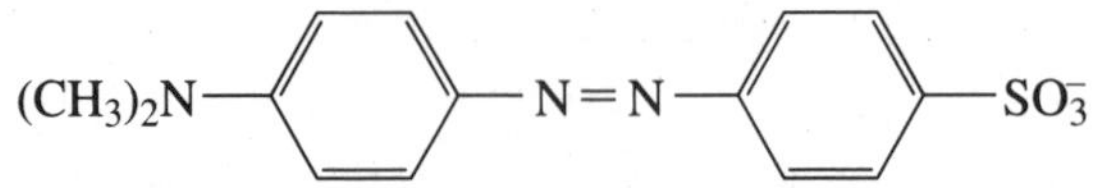

$$(CH_3)_2N-C_6H_4-N{=}N-C_6H_4-SO_3^-$$

Thus the ΔG° of the latter must be of the order of -1000 cal mol^{-1} or smaller.

From the stoichiometric binding constants in a series of homologous compounds, one can evaluate the contributions of structural substituents to the interaction energy of receptor with ligand (see Table 8.2). Serum albumin binds fatty acids as well as many other small molecules. In the series from C_2 to C_8 carboxylates, the affinities increase with chain length, although the observed ΔG° values vary also with buffer, pH, and species of albumin. The movement in affinity is about 900 cal mol^{-1} for each added CH_2 group. It does not follow, however, that the incremental affinity per CH_2 substituent would be the same for every base structure.

It is of interest to note that with larger ligands, such as oligosaccharide antigens, the ΔG° of binding with a specific monoclonal antibody does not change much with increasing length. The interaction free energy change stays in the range of 7 to 8 kcal mol^{-1} as the antigen varies in length from 3-mer to 20-mer.

8.2. CHANGE IN ΔG° IN SUCCESSIVE STOICHIOMETRIC STEPS

The trends in ΔG° as successive moles of ligand are bound may reflect a multitude of features in the interactions. In the anion–transferrin series (see Table 8.1), $\Delta G_2^\circ - \Delta G_1^\circ$ for the first to the second stoichiometric binding step is essentially identical for all three ligands (bicarbonate, phosphate, and vanadate). Evidently, there are no specific chemical interactions between the anions and the protein matrix, for these should be reflected by differences in $\Delta(\Delta G^\circ)$.

In some cases, it is possible to account quantitatively for the progressive changes in ΔG_i° in terms of clearly defined contributions, such as electrostatic interactions. For example, the list of K_i values for binding of the dianion azosulfathiazole by bovine serum albumin (34) (see Table 6.2) can be transformed into corresponding ΔG_i° values (Table 8.3). At each step i when one of the ligand dianions is bound

TABLE 8.3 Free Energy Changes and Electrostatic Contributions in Successive Steps in the Binding of Azosulfothiazole Dianion by Bovine Serum Albumin at 5°C

K_i	ΔG_i° (cal mol^{-1})	$\Delta(\Delta G^\circ_{\text{electrostatic}})$ (cal mol^{-1})
$K_1 = 1.25 \times 10^5$	−6483	154
$K_2 = 4.50 \times 10^4$	−5919	154
$K_3 = 2.16 \times 10^4$	−5514	154
$K_4 = 1.17 \times 10^4$	−5174	154
$K_5 = 6.70 \times 10^3$	−4867	154
$K_6 = 3.99 \times 10^3$	−4580	154
$K_7 = 2.43 \times 10^3$	−4306	154
$K_8 = 1.51 \times 10^3$	−4042	154
$K_9 = 9.5 \times 10^2$	−3785	154
$K_{10} = 6.0 \times 10^2$	−3530	154
$K_{11} = 3.8 \times 10^2$	−3280	154
$K_{12} = 2.4 \times 10^2$	−3030	154
$K_{13} = 1.54 \times 10^2$	−2780	154
$K_{14} = 7.0 \times 10^1$	−2525	154
$K_{15} = 4.6 \times 10^1$	−2268	154
$K_{16} = 2.5 \times 10^1$	−2004	154

to the receptor, the electrostatic field around the complex becomes more repulsive to a succeeding anion and hence the affinity $-\Delta G_i^\circ$ is decreased. This drop is in addition to the contribution from the statistical $(K_i/K_{i-1})_{\text{ideal}}$ factor reflecting the decrease in number of open binding sites on the ligand RL_i as i increases.

The statistical factor can be evaluated from equation 7.3:

$$\left(\frac{K_i}{K_{i-1}}\right)_{\text{ideal}} = \left(\frac{n-i+1}{n-i+2}\right)\left(\frac{i-1}{i}\right) \tag{7.3}$$

if one can make a reasonable guess for n. To obtain a measure of the electrostatic factor, first recognize the following equilibria:

$$K_i = \frac{(RL_i)}{(RL_{i-1})(L)} \tag{8.2}$$

$$K_{i-1} = \frac{(RL_{i-1})}{(RL_{i-2})(L)} \tag{8.3}$$

Consequently, K_i/K_{i-1} corresponds to the reaction

$$2RL_{i-1} = RL_i + RL_{i-2} \tag{8.4}$$

Each of the species in equation 8.4 carries a different electrical charge because the number of bound dianions varies. The corresponding electrostatic free energies can be calculated by well-known methods (34). A good fit to the experimental curve for B versus log (L) is obtained with an electrostatic stepwise $\Delta(\Delta G^\circ)$ (see Table 8.3) of 154 cal mol^{-1}. This value corresponds with what the electrostatic charging energies would be for a spherical receptor of about 30 Å in radius (34). Similar computations have been made for the binding of small monovalent anions such as Cl^- and SCN^- (1), which have greater charge densities. Here, too, the drop in affinity of the receptor with successive uptake of ligand ions can be accounted for by electrostatic interactions added to the statistical factors of decreasing number of open binding sites.

8.3. VARIATION OF ΔG° WITH THE NATURE OF THE RECEPTOR

In Table 8.2 we can discern examples of global effects and of local influences. The table summarizes, by ΔG° values, a set of interactions of cupric ion with serum albumin, β-lactoglobulin, and lysozyme respectively, in a particular buffer at a common pH. There is a striking parallelism between the affinity for cupric ion ligand and the isoelectric points of the respective proteins: serum albumin, 4.7; β-lactoglobulin, 5.2; and lysozyme, 11.0. This correlation suggests that electrostatic influences are involved. We can expect that the lower the isoelectric point, the greater the negative charge on the protein and hence the stronger the attraction for cationic cupric ion. The observed affinities fit this electrostatic trend.

With the availability of constructed mutagenesis techniques, it is possible to examine the effects of systematic substitution of amino acid residues in contact areas of ligand–receptor complexes. In this way it is hoped that the contribution of specific side chains to the free energy of binding can be established.

An example of such a study, the lysozyme–antibody interaction, is summarized in Table 8.2. The Fv fragment of the monocloncal antibody specific for hen egg white lysozyme was isolated, its complex with antigen crystallized, and the molecular structure resolved by X-ray diffraction. Thereafter, a set of mutant Fv fragments with specific substitutions in the heavy chain were constructed, and the energies of antigen–antibody interactions were determined. Interestingly, the values of ΔG° for the mutants were all about the same, but weaker than that for the wild type. Mutant 17, carrying a single amino acid substitution, showed a ΔG° of -8700 cal mol^{-1}, which is not significantly different from that for mutant 11 (-8600 cal mol^{-1}), with five amino acid substitutions.

In protein–DNA interactions, comparisons of Cro protein binding by sequence-nonspecific DNA and by sequence-specific DNA show distinct correlations with changes in sequence (see Table 8.2). The strongest interactions are with the specific OR3 and OR1 operators. The nonoperator and the nonspecific DNA receptors have ΔG° values several kilocalories weaker. Mutant DNA receptors with single substitutions at twofold related positions in the palindromic sequence almost always decrease the affinity for Cro protein by several kilocalories. The interpretation of these trends requires examination of the corresponding enthalpies and entropies of binding and considerations of the molecular forces involved in ligand–receptor interactions.

For most receptors, the affinity for the ligand does not change significantly with variations in receptor concentration; however, there are exceptions. For example, high concentrations of a negatively charged receptor such as the protein calbindin show much reduced affinities for the ligand cationic divalent calcium (35) (see Table 8.4). The binding constant for site 2 (see Fig. 2.9) when site 1 is occupied by calcium ion $k_{1,2}$ progressively drops 20-fold as the protein concentration is increased about 200-fold (see Table 8.4). This trend is a manifestation

TABLE 8.4 Dependence of Affinity on Concentration of Receptor

Concentration of Calbindin (mM)	Concentration of Sodium ion (mM)	$k_{1,2}$
0.027	0.14	4.5×10^5
0.097	0.5	3.0×10^5
0.380	2.0	1.9×10^5
0.850	4.4	1.3×10^5
2.63	13.7	0.55×10^5
7.35	38.2	0.26×10^5

of changes in electrostatic screening due to increases in ionic strength in the solution upon addition of the charged protein and its sodium counterion. These changes can be accounted for by Monte Carlo simulations that calculate the screening effects quantitatively; good agreement has been found between theory and experimental observations (35). In principle, similar screening effects should be present with other charged proteins and receptors, but they may be obscured by other electrolytes in solution as well as by conformational accommodations in the macromolecule.

8.4. ENTHALPIES AND ENTROPIES OF BINDING

In most investigations of the energetics of ligand–receptor binding, the enthalpy of reaction is obtained from studies of the variation of affinity with temperature. When the temperature dependence of K_i has been ascertained, one can obtain ΔH° from the relation (36)

$$\frac{d \ln K}{dT} = \frac{\Delta H^\circ}{RT^2} \tag{8.5}$$

often called the van't Hoff equation. In practice, this equation is rearranged into

$$d \ln K = -\frac{\Delta H^\circ}{R} d\left(\frac{1}{T}\right) \tag{8.6}$$

which suggests that one should plot log K_i versus $1/T$ and compute ΔH° from the slope of the curve.

There is a widespread misconception that such a van't Hoff plot must always be linear and that the slope ΔH° must be a constant, often denoted $\Delta H^\circ_{\mathrm{vH}}$, the van't Hoff heat of reaction. This unwarranted presumption can lead to wildly incorrect conclusions.

An illustration of the pitfalls that can be encountered is provided in a recent study of the binding of cytidine 2′-monophosphate (2′-CMP) by ribonuclease A (RNase A) (14). A plot of log K versus $1/T$ is presented in Figure 8.1. A possible straight line through the data is also shown. In view of the uncertainties in the values of the K (even on the logarithmic scale of the ordinate axis), the line shown seems

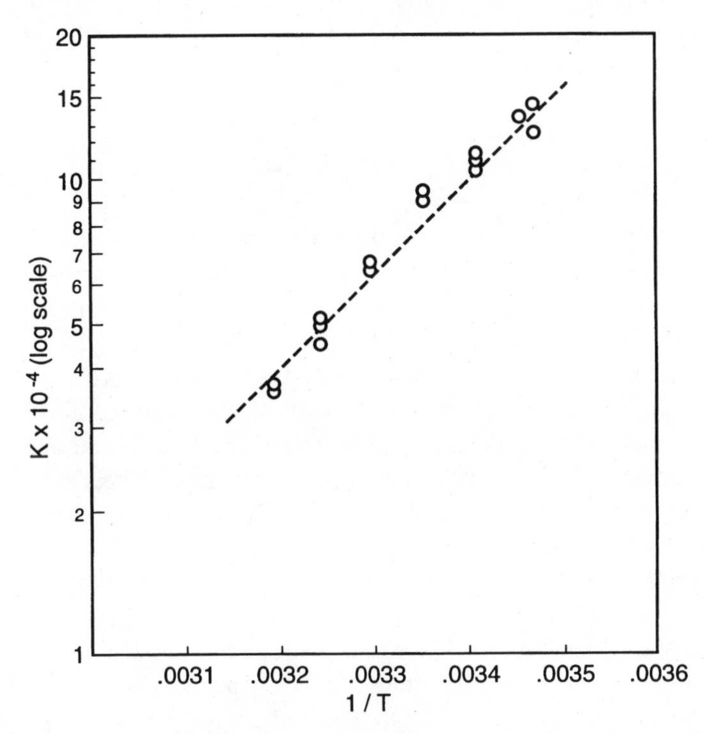

Figure 8.1 The van't Hoff plot of equilibrium constants for the binding of cytidine 2′-monophosphate by ribonuclease A. A possible straight line through the data is illustrated.

reasonable to the eye. This assessment is confirmed if one carries out a linear least squares analysis of the graph, which gives a correlation coefficient of 0.979 (14) and a constant slope $\Delta H^\circ_{\mathrm{vH}}$ of −9.47 kcal mol^{-1}. In fact, however, when the ΔH° values of the binding of 2′-CMP by RNase A are measured calorimetrically ΔH° is found to vary from −8.82 to −13.99 kcal mol^{-1} over the temperature range of 15 to 40°C, the span covered in Figure 8.1 (14). Obviously, ΔH° is not independent of temperature. In fact, if you look at Figure 8.1 carefully, ignoring the straight line, the points shown, despite their uncertainties, sketch out a curve, with the slope ($-\Delta H^\circ$) at low temperature (high $1/T$) being appreciably smaller than that at high temperature (low $1/T$).

Enthalpies of binding can also be measured directly, as has become increasingly common with the commercial availability of microcalorimeters.

Once ΔH° has been evaluated, entropy changes can be computed from the thermodynamic relation (36)

$$\Delta G^\circ = \Delta H^\circ - T\Delta S^\circ \tag{8.7}$$

Table 8.2 presents values of enthalpies and entropies of binding for a wide range of ligands and of receptors. It might be expected that a favorable ΔG°, a negative number, would be associated with a negative value for ΔH°, corresponding to a drop in internal energy ($\Delta E \simeq \Delta H$) of the complex compared with the separated constituents. Although equation 8.7 draws attention to the additional contributions of ΔS° to ΔG°, it is expected that it would be unfavorable; the free ligand and receptor presumably should have more degrees of motional freedom than the ligand–receptor complex. As Table 8.2 shows, however, the observed thermodynamic values do not fit such a simple pattern.

There are many ligand–receptor complexes for which ΔH° is negative and larger numerically than the corresponding ΔG° (e.g., cytidinephosphate–ribonuclease and lysozyme–antibody Fv). In these instances, ΔS° is a negative number, as expected in a simple view of two combining species. On the other hand, there are many examples of ligand–receptor complexes for which ΔH° is negative but much smaller numerically than ΔG° (e.g., hexadecylphosphorylcholine–phospholipase, caproate–albumin, and oligosaccharide 3-mer–antibody) or near zero (e.g., chloride– and thiocyanate–albumin) or even a positive number (e.g., cupric ion–albumin and Cro protein–DNA). In the last circumstance, the complex has a higher internal energy than does separated ligand and receptor, so that ΔE° ($\simeq \Delta H^\circ$) must be positive, i.e., E° goes uphill. When a negative ΔG° is not due to a dominant favorable ΔH°, then the complexation is driven by a *positive* entropy change (see Table 8.2), a surprising result for two separate species combining to form a complex. For insight into this enigma, the next chapter turns to an examination of the forces involved in ligand–receptor interactions.

REFERENCES

1. G. Scatchard, I. H. Scheinberg, and S. H. Armstrong Jr., *J. Am. Chem. Soc.*, **72**, 535–540 (1950).
2. I. M. Klotz and H. G. Curme, *J. Am. Chem. Soc.*, **70**, 939–943 (1948).

3. I. M. Klotz, in H. Neurath and K. Bailey, eds., *The Proteins*, New York, Academic Press, Inc., 1953, Chapt. 8.
4. J. D. Teresi and J. M. Luck, *J. Biol. Chem.*, **194**, 823–834 (1952).
5. B. Honoré and R. Brodersen, *Anal. Biochem.*, **171**, 55–66 (1988).
6. I. M. Klotz and J. M. Urquhart, *J. Am. Chem. Soc.*, **71**, 847–851 (1949).
7. I. M. Klotz, *Cold Spring Harb. Symp. Quant. Biol.*, **14**, 97–112 (1950).
8. B. Honoré and R. Brodersen, *Mol. Pharmacol.*, **25**, 137–150 (1984).
9. R. Brodersen, B. Honoré, A. O. Pedersen, and I. M. Klotz, *Trends Pharmacol. Sci.*, **9**, 252–257 (1988).
10. A. Knudsen, A. O. Pedersen, and R. Brodersen, *Arch. Biochem. Biophys.*, **244**, 273–284 (1986).
11. R. Brodersen and A. Robertson, *Mol. Pharmacol.*, **36**, 478–483 (1989).
12. P. Soares de Araujo, M. Y. Rosseneu, J. M. H. Kremer, E. J. J. van Zoelen, and G. H. de Haas, *Biochemistry*, **18**, 580–586 (1979).
13. M. C. Chervenak and E. J. Toone, *J. Am. Chem. Soc.*, **116**, 10533–10539 (1994).
14. H. Naghibi, A. Tamura, and J. M. Sturtevant, *Proc. Natl. Acad. Sci. USA*, **92**, 5597–5599 (1995).
15. N. M. Allewell, J. Friedland, and K. Niekamp, *Biochemistry*, **14**, 224–230 (1975).
16. S. F. Velick, *Fed. Proc.*, **10**, 264 (1951).
17. P. R. Connelly, J. A. Thomson, M. J. Fitzgibbon, and F. J. Bruzzese, *Biochemistry*, **32**, 5583–5590 (1993).
18. J. F. Halsey and R. L. Biltonen, *Biochemistry*, **14**, 800–804 (1975).
19. L. Soltés, B. Sébille, and A. Fügedi, *Bio-Sciences*, **8**, 13–17 (1989).
20. L. Soltés, B. Sébille, and P. Szalay, *J. Pharm. Biomed. Anal.*, **12**, 1295–1302 (1994).
21. B. L. Jacobsen and F. A. Quiocho, *J. Mol. Biol.*, **204**, 783–787 (1988).
22. B. W. Sigurskjold, E. Altman, and D. R. Bundle, *Eur. J. Biochem.*, **197**, 239–246 (1991).
23. W. Ito, Y. Iba, and Y. Kurosawa, *J. Biol. Chem.*, **268**, 16639–16647 (1993).
24. G. Rialdi, S. Raffanti, and F. Manca, *Mol. Immunol.*, **21**, 945–948 (1984).
25. A. D. Johnson, A. R. Poteete, G. Lauer, R. T. Sauer, G. K. Ackers, and M. Ptashne, *Nature*, **294**, 217–223 (1981).
26. M. Ptashne, *A Genetic Switch*, 2nd ed., Blackwell Scientific Publications and Cell Press, Cambridge, Mass., 1992.
27. Y. Takeda, P. D. Ross, and C. P. Mudd, *Proc. Natl. Acad. Sci. USA*, **89**, 8180–8184 (1992).

28. T. Clackson, and J. A. Wells, *Science*, **267**, 383–386 (1995).
29. I. M. Klotz and J. M. Urquhart, *J. Phys. Chem.*, **53**, 100–114 (1949).
30. K. Landsteiner and J. van der Scheer, *J. Exp. Med.*, **48**, 315–325 (1928).
31. F. Karush, *J. Phys. Chem.*, **56**, 70–78 (1952).
32. K. Landsteiner, *The Specificity of Serological Reactions*, Harvard University Press, Cambridge, Mass. 1945.
33. H. N. Eisen and F. Karush, *J. Am. Chem. Soc.*, **71**, 363–368 (1949).
34. I. M. Klotz, F. M. Walker, and R. B. Pivan, *J. Am. Chem. Soc.*, **68**, 1486–1490 (1946).
35. S. Linse, B. Jönson, and W. J. Chazin, *Proc. Natl. Acad. Sci. USA*, **92**, 4748–4752 (1995).
36. I. M. Klotz and R. M. Rosenberg, *Chemical Thermodynamics*, 5th ed., John Wiley & Sons, Inc., New York, 1994.

CHAPTER 9

FORCES OF INTERACTION

Insights into forces of molecular interactions are derived largely from observations of the behavior of atoms and molecules in the gas phase, coupled with measurements on pure liquids and pure solids. From such studies several categories of interaction can be discerned.

9.1. NONPOLAR ATTRACTIONS: VAN DER WAALS, LONDON, HYDROPHOBIC

Two molecules for which energies of transformation from the gas state to a condensed phase are known are

$$\mathrm{Ar(g)} = \mathrm{Ar(s)} \qquad \Delta H^\circ = -2\ \mathrm{kcal\ mol^{-1}} \tag{9.1}$$

$$\mathrm{C_4H_{10}(g)} = \mathrm{C_4H_{10}}(l) \qquad \Delta H^\circ = -5\ \mathrm{kcal\ mol^{-1}} \tag{9.2}$$

In both cases there is a drop in internal energy ($\Delta H^\circ \simeq \Delta E^\circ$) in the conversion of widely separated atoms or molecules into closely packed interacting molecules in the solid or liquid state. Because of different electronegativities of C and H atoms, there is some separation of charge in the C—H bond and hence a dipole is created. Dipole–dipole interactions between nonpolar molecules thus become possible, as was recognized by van der Waals over a century ago. In contrast, argon is a spherical, symmetric atom, with no permanent dipole. Nevertheless, it can interact with other argon atoms, for reasons discovered by London after the advent of wave mechanics. London forces can also supplement, and be coupled with, van der Waals interactions.

In aqueous solutions, van der Waals and London forces operate between solute and solvent molecules as well as between molecules of each component. The net change in ΔH° (or ΔE°) thus depends on the relative contributions of the different energies. Even if the net result is a positive ΔH°, unfavorable to the occurrence of a transformation, the chemical or physical reaction is often observed to proceed. Clearly that means that ΔS° must be large and positive so that the $-T\Delta S^\circ$ term in equation 8.7 will overwhelm the unfavorable ΔH° term to give a favorable, i.e., negative value, for ΔG°. But how can ΔS° be a positive number for a transformation in which two separated species (eg, ligand and receptor) *combine* to form a complex?

To gain some insight into this enigma, let us examine the transformation

$$
\begin{aligned}
C_3H_8(\mathrm{aq}) &= C_3H_8(l) \qquad (9.3)\\
\Delta G^\circ &= -4 \text{ kcal mol}^{-1}\\
\Delta H^\circ &= +2 \text{ kcal mol}^{-1}\\
\Delta S^\circ &= +20 \text{ cal mol}^{-1}\text{deg}^{-1}\\
-T\Delta S^\circ &= -6 \text{ kcal mol}^{-1}
\end{aligned}
$$

In line with long experience that propane dissolved in water at standard concentration will move spontaneously to an adjacent phase of liquid propane, we find that the free energy change ΔG° is a substantial negative number. But the transfer is energetically *uphill*, $\Delta H^\circ \simeq \Delta E^\circ = +2$ kcal mol^{-1}. So ΔS° must be, and is, positive.

This puzzle was recognized by theoretical chemists and physicists over half a century ago. By the mid-1930s they realized that the puzzle could be resolved if it were assumed that nonpolar solutes in aqueous solution restricted the orientations and motions of the solvent water molecules. This concept was strengthened and elaborated on and served to explain a wide range of observations (1, 2).

For example, in regard to the transfer depicted in equation 9.3, we visualize the insertion of the inert molecule of C_3H_8 in aqueous solution as spontaneously triggering the rearrangement of H_2O molecules of the solvent into a more orderly framework or sheath around the solute molecule. In this hydration shell, hydrogen bonding is stronger than in the ambient pure solvent. That strengthening in itself increases the separation between energy levels and decreases the entropy. Furthermore, it stiffens the constraints on the orientational motions of the constituent water molecules. In consequence, there is an additional decrease in entropy of the system.

Thus when a C_3H_8 molecule is transferred out of the aqueous solvent into the pure propane liquid phase, the C_3H_8 releases the hydration sheath and the liberated H_2O molecules become dispersed in the aqueous solvent, where they have substantially increased orientational options. Consequently, the entropy of the system increases as C_3H_8 in aqueous solution moves into the pure liquid propane, i.e., ΔS° is positive for the transfer depicted in equation 9.3. In the reverse direction, when C_3H_8 moves from liquid propane to aqueous solution, ΔS° is negative because the H_2O molecules form a constrained sheath around the solute molecule.

Likewise, two separated nonpolar molecules M and M, in aqueous solution will experience an attraction for each other because the formation of the complex M_2 would be accompanied by a positive ΔS° arising from the loosening and release of some of the water molecules in the hydration sheaths around the individual, independent nonpolar M molecules. In anthropomorphic language, two nonpolar molecules in aqueous solution are attracted to each other by an entropically powered force. This is the defining feature of a hydrophobic bond. This is the characteristic that distinguishes a hydrophobic bond from a van der Waals or London interaction.

9.2. IONIC AND ELECTROSTATIC INTERACTIONS

In the gas phase, a cation and an anion are strongly attracted to each other by electrostatic forces:

$$Na^+(g) + Cl^-(g) = NaCl(s) \qquad \Delta H^\circ = -187 \text{ kcal mol}^{-1} \tag{9.4}$$

As the ΔH° value shows, there is an enormous drop in energy, 187,000 cal mol^{-1} when sodium ions combine with chloride ions. On the other hand, these same ions in aqueous solution remain almost entirely separated:

$$\mathrm{Na^+(aq) + Cl^-(aq) = NaCl(s) + water} \qquad \Delta G^\circ > 0; \Delta H^\circ \simeq 0 \quad (9.5)$$

Until Arrhenius marshaled evidence for this viewpoint, it was assumed essentially universally that sodium chloride in aqueous solution must exist as neutral NaCl molecules, for the separated neutral atoms would react almost explosively with water, and separated Na^+ and Cl^- species should experience an enormous attractive force between them (see equation 9.4). Arrhenius emphasized, however, that in water there would be a very large energy change due to ion–solvent interactions that could exceed the energy of electrostatic attractions. This view was overwhelmingly confirmed, and in time, quantitative values for ion-hydration interactions (ie, ΔH° for transfer of the ion in the gas phase to water solution) became available. Some representative values are shown in Table 9.1.

It should also be noted that in Table 9.1 the ionic entropies of hydration are all negative numbers. That is what one would expect if there were strong electrostatic interactions between the charged solute ion and dipolar H/O\H molecules of water.

With this background, it comes as no surprise to learn that when two ionic species in aqueous solution do combine, the association is

TABLE 9.1 Hydration Energies

Ion or Molecule	ΔH° hydration (kcal mol^{-1})	ΔS° (cal mol^{-1} deg^{-1})
H^+	−258	−26
Na^+	−95	−28
K^+	−75	−12
Ca^{++}	−337	−50
Al^{+++}	−1109	−111
OH^-	−111	—
Cl^-	−90	−24
Br^-	−82	−20
Ar	−3	—
CH_4	−4	—
C_2H_6	−5	—
C_3H_8	−6	—
C_4H_{10}	−5	—
C_2H_5Cl	−8	—

accompanied by a positive ΔS°. For example, for

$$H^+(aq) + Ac^-(aq) = HAc(aq) \qquad \Delta S^\circ = 22 \text{ cal mol}^{-1}\text{deg}^{-1}$$
$$\Delta H^\circ \simeq 0 \tag{9.6}$$
$$\Delta G^\circ = -6.5 \text{ kcal mol}^{-1}$$

$$H^+(aq) + SO_4^=(aq) = HSO_4^- \qquad \Delta S^\circ = 27 \text{ cal mol}^{-1}\text{deg}^{-1}$$
$$\Delta H^\circ = +5.2 \text{ kcal}^{-1}\text{mol}^{-1} \tag{9.7}$$
$$\Delta G^\circ = -2.7 \text{ kcal}^{-1}\text{mol}^{-1}$$

In both cases, the spontaneous ion–ion combinations, with favorable values for ΔG°, are entropy-driven reactions. At the molecular level, this is eminently reasonable, despite the fact that two independent species form a complex, for this combination must be accompanied by a release of some water molecules from the hydration sheaths of the separated ions and the consequent loosening of motional constraints on these solvent molecules.

These features are not limited to combinations of H^+ ion with bases. Similar thermodynamic characteristics are evident in complexation reactions of metallic ions. A selected set is listed in Table 9.2. These examples show that the formation of coordination complexes may also be driven by an entropically powered force.

In addition to combinations between ionic species, other electrostatic interactions are also encountered, such as ion–dipole, dipole–dipole, and ion- or dipole-induced dipole. Often overlooked are quadrupolar interactions, which can involve molecules that are gen-

TABLE 9.2 Thermodynamic Properties of Formation of Coordination Complexes

Coordinating Species	ΔG° (cal mol^{-1})	ΔH° (cal mol^{-1})	ΔS° (cal mol^{-1} deg^{-1})
Ca^{2+}; SO_4^{2-}	−3,200	1,700	16
Fe^{3+}; SO_4^{2-}	−5,700	6,200	39
Fe^{3+}; Cl^-	−2,000	8,500	35
Fe^{3+}; F^-	−6,700	7,500	49
Mg^{2+}; $EDTA^{4-}$	−11,700	3,500	51
La^{3+}; $EDTA^{4-}$	−20.700	−2,900	60

erally viewed as nonpolar or hydrophobic. Benzene, for example, has a π-electron cloud above the plane of C_6H_6 and an equivalent one below (Fig. 9.1A). As a result, a quadrupole moment is present. The interaction between a point charge and a quadrupole can be strong. For example, in the gas phase, the $K^+—C_6H_6$ binding energy is 19 kcal/mol, compared with 18 kcal/mol for $K^+—H_2O$ (3).

Even polar, nonionic molecules, such as H_2O and NH_3, form complexes with C_6H_6 in the gas phase (4, 5). As is shown in Figure 9.1B, an NH_3 residing above the C_6H_6 plane supplies an N—H proton that interacts with the π-electron cloud of benzene. In the complex of benzene with water, both protons of $\mathrm{H}{\diagup}^{\mathrm{O}}{\diagdown}\mathrm{H}$ are directed toward and interact with the π-electron cloud of the ring.

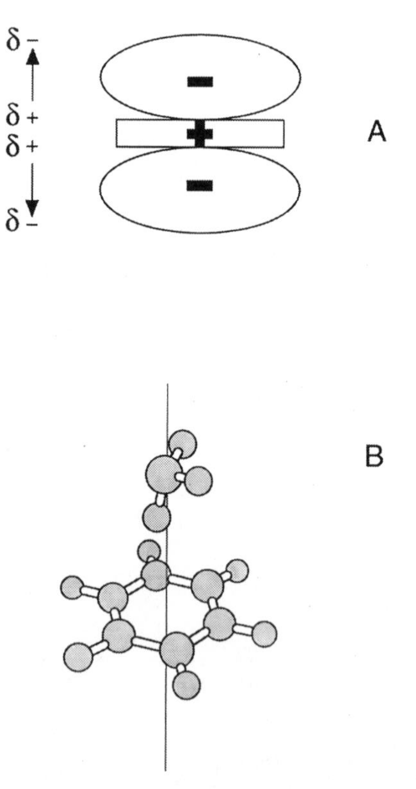

Figure 9.1 **A,** Schematic representation of the π-electron clouds above and below the plane of a benzene ring. The separation of partial charges creates a quadrupole moment as shown by the opposing dipolar arrows. **B,** Orientation of NH_3 and C_6H_6 constituents of the benzene–ammonia complex.

It has also been shown that cation–quadrupole interactions in aqueous solutions can be as strong as those between an ion and a dipolar molecule such as H_2O (6). Suitably designed aromatic molecules providing a ring of quadrupoles can attract a cation out of an aqueous environment into a nonpolar one. There is good reason to expect that a cationic side chain in a protein, such as that from a lysine or arginine residue, would interact with the quadrupoles of phenylalanine, tyrosine, or tryptophan.

It can be shown by a simple theoretical argument that in all electrostatic interactions—not just ion–ion but also in ion–dipole, dipole–dipole, etc.—the free energy change should have a large favorable contribution from the entropy component (7).

If the favorable free energy of formation of a complex from its constituent species is due to interactions that are electrostatic in origin, then ΔG_{elect} would be a function of certain atomic dimensions r, the charges z on the species involved, and the dielectric constant D of the medium (8–10). By associating all of the parameters except D in a single function ϕ we may write

$$\Delta G_{\text{elect}} = \phi(r,z)/D \tag{9.8}$$

If we made the additional reasonable assumption that r and z are independent of the temperature, it follows that

$$\frac{\partial \Delta G_{\text{elect}}}{\partial T} = \phi(r,z)\,\frac{\partial(1/D)}{\partial T} = -\frac{\phi(r,z)}{D^2}\,\frac{\partial D}{\partial T}. \tag{9.9}$$

This equation may be rearranged to

$$\frac{\partial \Delta G_{\text{elect}}}{\partial T} = \frac{-\phi(r,z)}{D}\,\frac{\partial \ln D}{\partial T} = -\Delta G_{\text{elect}}\,\frac{\partial \ln D}{\partial T}. \tag{9.10}$$

If we use the dielectric constant of pure water for D, the temperature coefficient may be obtained from the following equation (11):

$$\log_{10} D = 1.9446 - 0.00198t \tag{9.11}$$

where t is temperature in °C. Equation 9.10 thus becomes

$$\frac{\partial \Delta G_{\text{elect}}}{\partial T} = +0.00456 \Delta G_{\text{elect}} \tag{9.12}$$

By recognizing that the temperature coefficient of the free energy is the negative of the entropy, we obtain

$$\Delta S_{\text{elect}} = -0.00456 \Delta G_{\text{elect}} \tag{9.13}$$

or

$$-T\Delta S_{\text{elect}} = 0.00456T(\Delta G_{\text{elect}}). \tag{9.14}$$

For a temperature near 298°K,

$$-T\Delta S_{\text{elect}} = 1.35 \Delta G_{\text{elect}} \tag{9.15}$$

Finally, from the general thermodynamic expression

$$\Delta G = \Delta H - T\Delta S \tag{9.16}$$

we find the following equation for ΔH_{elect}:

$$\Delta H_{\text{elect}} = \Delta G_{\text{elect}}(1 - 0.00456T) \tag{9.17}$$

At $T = 298$°K,

$$\Delta H_{\text{elect}} = -0.35 \Delta G_{\text{elect}} \tag{9.18}$$

Thus we see from equation 9.15 that $-T\Delta S_{\text{elect}}$ is of the order of magnitude of ΔG_{elect}, ie, is the dominant contributor to the free energy change, whereas ΔH_{elect} (equation 9.18), contributes only a small portion. Clearly, then, all types of electrostatic bonding should also be entropically driven. *Positive entropies for binding of ligands by receptors (Table 8.2) are thus not diagnostic of hydrophobic interactions but may be instead a manifestation of electrostatic effects.*

9.3. HYDROGEN BONDING

The concept of the hydrogen bond was discovered or invented by Huggins in 1919 while he was still a graduate student, and was brought to general attention a year later by two of his older colleagues (12).

Early estimates of the stability of the hydrogen bond were obtained from studies in the vapor phase. Association equilibria of carboxylic acids were examined first by classical measurements of deviations from ideal gas behavior and subsequently were refined and extended by molecular spectroscopy studies. With these techniques, the association of molecules such as acetic acid

$$2CH_3{-}C(=O){-}O{-}H\,(g) \rightleftharpoons CH_3{-}C\begin{matrix} /\!\!/ O\cdots H{-}O \backslash \\ \backslash O{-}H\cdots O /\!\!/ \end{matrix}C{-}CH_3\,(g) \qquad (9.19)$$

was examined and found to be accompanied by a ΔH° near -14 kcal mol^{-1} dimer. Since the dimer is held together by two hydrogen bonds, the ΔH° for formation of an O···H—O in the gas phase from the separated constituents must be -7 kcal/mol (Table 9.3).

To be more relevant to receptors, let us focus on proteins. The major source of hydrogen bonding in these macromolecules are the amide hydrogen bonds:

TABLE 9.3 Thermodynamic Properties for Formation of Hydrogen Bonds

Molecule(s) involved	State	Type of Bond	ΔH° (kcal mol^{-1})	ΔS° cal mol^{-1} deg^{-1}
H—COOH[a]	Gas	O···H—O	−7.4	−18
CH_3—COOH[a]	Gas	O···H—O	−7.3	−18
CH_3—OH[a]	Gas	O···H—O	−7.6	−23
CH_3—OH/$N(CH_3)_3$[a]	Gas	N···H—O	−5.9	−22
CH_3—OH/$N(C_2H_5)_3$[a]	Gas	N···H—O	−7.6	—
$CH_3{-}C(=O){-}N(H){-}CH_3$ dissolved in:[b]	CCl_4	C=O···H—N	−4.2	−11
	Trans-ClHC=CHCl	C=O···H—N	−3.3	−10
	Cis-ClHC=CHCl	C=O···H—N	−1.5	−5
	Dioxane	C=O···H—N	−0.8	−4
	Acetonitrile	C=O···H—N	−0.7	−3
	Water	C=O···H—N	−0.0	−10

[a]Ref. 13.
[b]Refs. 14–16.

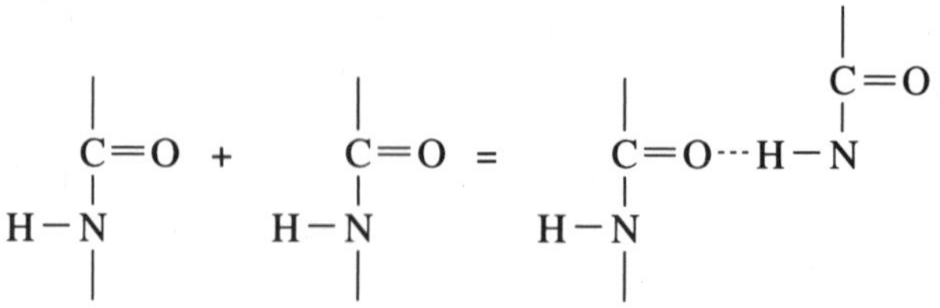

$$
\begin{array}{c} | \\ C{=}O \\ | \\ H{-}N \\ | \end{array}
\; + \;
\begin{array}{c} | \\ C{=}O \\ | \\ H{-}N \\ | \end{array}
\; = \;
\begin{array}{c} | \\ C{=}O \\ | \\ H{-}N \\ | \end{array}
\!\cdots H{-}\begin{array}{c} | \\ C{=}O \\ | \\ N \\ | \end{array}
\tag{9.20}
$$

One might estimate the strength of this C=O···H—N bond from the ΔH° values in the gas phase for the analogous hydrogen bonds listed in Table 9.3. However if we are focusing on proteins, such numbers are not really pertinent. What we need to know for a peptide hydrogen bond that might be involved in stabilizing a protein conformation is its stability in an aqueous environment. In solution in water the interpeptide hydrogen bond is exposed to competing bonds with water. So the following interchange can occur:

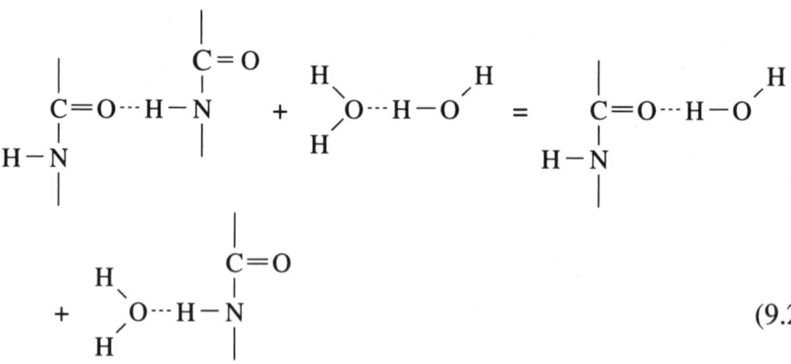

$$
\begin{array}{c} | \\ C{=}O \\ | \\ H{-}N \\ | \end{array}
\!\cdots H{-}\begin{array}{c} | \\ C{=}O \\ | \\ N \\ | \end{array}
\; + \;
\begin{array}{c} H \\ \diagdown \\ O \\ \diagup \\ H \end{array}\!\cdots H{-}O\!\begin{array}{c} H \\ \diagup \end{array}
\; = \;
\begin{array}{c} | \\ C{=}O \\ | \\ H{-}N \\ | \end{array}
\!\cdots H{-}O\!\begin{array}{c} H \\ \diagup \end{array}
$$

$$
+ \;
\begin{array}{c} H \\ \diagdown \\ O \\ \diagup \\ H \end{array}\!\cdots H{-}\begin{array}{c} | \\ C{=}O \\ | \\ N \\ | \end{array}
\tag{9.21}
$$

The net energy change, or ΔH°, for the bond disruption in aqueous solution would be expected to be much smaller than that for the opening of the N—H···O=C bond in the gas phase or in a nonpolar, nonhydrogen bonding solvent.

Since in a protein each of the open bonds in the naked entity $-\overset{\displaystyle O}{\overset{\|}{C}}-\underset{\displaystyle H}{\underset{|}{N}}-$ is attached to a carbon atom of the polypeptide chain, *N*-methylacetamide $H_3C-\overset{\displaystyle O}{\overset{\|}{C}}-\underset{\displaystyle H}{\underset{|}{N}}-CH_3$ can serve as a prototype of the peptide group. As Table 9.3 illustrates, the ΔH° of formation of the interpeptide hydrogen bond of *N*-methylacetamide decreases (in absolute value) in nonaqueous solvents, as one goes from nonpolar to increasingly polar solvents. The trend from carbon tetrachloride to acetonitrile points to a near zero value for ΔH° in water, and that is indeed what has been found experimentally. These results show clearly that the amide hydrogen bond energy in an aqueous environment is near zero.

There is an inclination offhand to think that if the C═O···H—N bond is within a nonpolar hydrophobic pocket within a protein structure its bond energy would be nearer 7 kcal mol^{-1}. A ΔH° value in that range might be pertinent for the C═O···H—N in the nonpolar (hydrophobic) pocket opening to expose freely rotating C═O and H—N groups to the same nonpolar pocket. However, protein breathing motions permit water molecules to enter (and leave) the interior of the macromolecule, and these would form the same aqueous hydrogen bonds displayed in equation 9.21. Thus the *equilibrium* state would involve hydrated C═O and H—N groups, and the ΔH° for producing these should be near zero.

These energetic relationships have been clearly resolved in some calorimetric experiments with *N*-methylacetamide in various states (17). The relative enthalpies (or internal energies) are presented in Figure 9.2. From this energy level diagram, based on direct calorimetric measurements, it can be concluded that for

$$\text{N—H} \cdots \text{O═C (nonpolar pocket)} = \text{N—H/O═C (aqueous)} \tag{9.22}$$

ΔH° is −2.7 kcal mol^{-1}. Thus the N—H···O═C bond in a nonpolar (hydrophobic) pocket is not an intrinsically stable hydrogen bond in the presence of water.

Despite the intrinsic instability of hydrogen bonds in model peptide molecules, there is no doubt that intramolecular hydrogen bonding exists in proteins even in aqueous solution. Once again, entropy factors play an important role. In a polypeptide, the amide groups are in close proximity so their local concentrations are high. To bring them into juxtaposition, covalent bonds had to be formed, either during cellular biosynthesis or during laboratory synthesis. In either event, some free energy (from some high energy compound) had to be expended to overcome the unfavorable $T\Delta S$ for bringing amide groups close together. Simple calculations show that the local concentration of peptide groups within the molecular volume of proteins is 12 to 13 molar

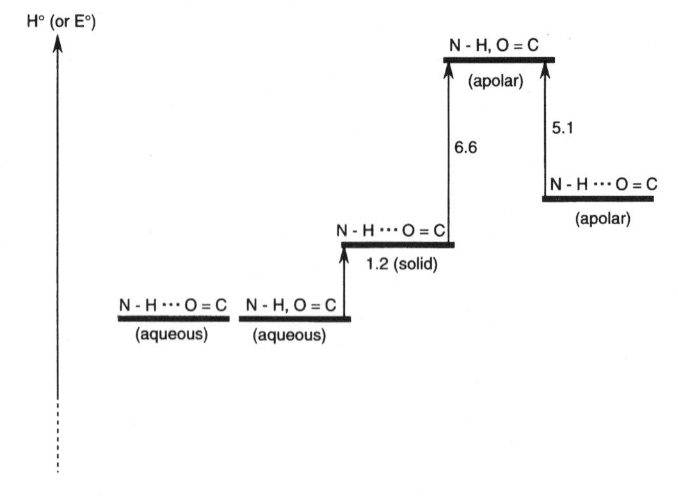

Figure 9.2 Relative enthalpies (or internal energies) for interpeptide hydrogen bond and its separated constituents as determined by calorimetric measurements. The actual experiments used the prototypical molecule $CH_3{-}C({=}O){-}N(H){-}CH_3$ (17).

---

(18). The prototypical molecule *N*-methylacetamide in a 12-molar solution is largely in the amide-hydrogen-bonded state.

Complicating hydrogen bonding even further is the observation that H_2O may play a key structural role in providing hydrogen bonds as bridges between a ligand and its receptor (19). For example, in the binding of cyclic urea inhibitors to human immunodeficiency virus protease, one H_2O molecule strategically placed in a void between ligand and receptor forms four hydrogen bonds, two to each constituent of the complex, without which it would not be stable.

9.4. CLATHRATE HYDRATES

There is still another way in which water may play a role in protein interactions, intramolecularly or in complexes with ligands. Water molecules are extraordinarily versatile in the kinds of structures that they can form or participate in (20). Under pressure, pure water can exist in 11 solid ices, ice VII being stable even above 50°C. Each has a different arrangement and packing density of hydrogen-bonded H_2O molecules.

It is also possible to obtain "ices" from water to which a small amount of impurity has been added. These are called polyhedral clathrate hydrates. The first one isolated, chlorine hydrate, was discovered by at the beginning of the nineteenth century (21, 22). Subsequently, the inert gases argon, krypton, and xenon were found to form crystalline stoichiometric hydrates, and by now more than a hundred polyhedral hydrates have been isolated (Table 9.4). In composi-

TABLE 9.4 Polyhedral Clathrate Hydrates

Class I		Class II	Class III
		Guest Molecules	
Ar	CH_4	$CHCl_3$	$(n\text{-}C_4H_9)_4N^+F^-$
Kr	C_2H_2	CH_3CHCl_2	$(n\text{-}C_4H_9)_4N^{+-}O_2CC_6H_5$
Cl_2	C_2H_4	$(CH_3)_2O$	$[(n\text{-}C_4H_9)_4N^+]_2WO_4^{2-}$
H_2S	C_2H_6	C_3H_8	$(i\text{-}C_5H_{11})_4N^+F^-$
PH_3	CH_3Cl	$(CH_3)_3CH$	$(n\text{-}C_4H_9)_3S^+F^-$
SO_2	CH_3SH	C_3H_7Br	$(n\text{-}C_4H_9)_4P^+Cl^-$
$C_2H_5NH_2$	CH_3CHF_2	$(CH_3)_2CO$	$(CH_3)_3N$ $(CH_2)_6N_4$
$(CH_3)_2NH$	CH_2CHF		$n\text{-}C_3H_7NH_2$
			$(CH_3)_4N^+OH^-$
$\overline{CH_2CH_2O}$ (ring)		C_6H_6	$i\text{-}C_3H_7NH_2$
			$(C_2H_5)_2NH$
$\overline{CH_2CH_2CH_2O}$ (ring)		$\overline{CH_2CH_2CH_2CH_2O}$ (ring)	$(CH_3)_3C—NH_2$
			C_4H_9OH
		cyclo-C_6H_{12}	
		$\overline{CH_2CH_2OCH_2CH_2O}$ (ring)	
		Stoichiometry	
$M \cdot 5\frac{3}{4}\ H_2O$		$M \cdot 17H_2O$	$M \cdot (5-40)H_2O$
$M \cdot 7\frac{2}{3}\ H_2O$		$M \cdot M'_2 \cdot 17H_2O$	
		Unit Cell	
$46H_2O$		$136H_2O$	Variable number of H_2O
Polyhedra[a]		Polyhedra[a]	Polyhedra[a]
2 H_{12}(5 Å diam.)		16 H_{12}(5 Å diam.)	H_8, H_{12}, H_{14}, H_{15}
6 H_{14}(6 Å diam.)		8 H_{16}(7 Å diam.)	H_{16}, H_{17}, H_{18}, H_{60}, etc.
Faces		Faces	Faces
Pentagons		Pentagons	Quadrilaterals
Hexagons		Hexagons	Hexagons
			Heptagons
$M_2 \cdot M_6 \cdot 46H_2O$		$M_8 \cdot M'_{16} \cdot 136H_2O$	

[a] H_n symbolizes a polyhedron with n faces.

tion, they are overwhelmingly water. X-ray crystallography has shown that the impurity, or guest molecule, is enclosed in a cage of water molecules. A few of the many different known cages are shown in Figure 9.3 (18, 23). It is apparent that many different polyhedra can be formed, with pentagonal (and even quadrilateral) as well as hexagonal faces, and with cavities for the enclosure of a wide range of sizes of guest molecules.

Among the molecules that form polyhedral hydrates are methane, propane, isobutane, methyl mercaptan, dimethyl sulfide, benzene, and even a cluster of four isoamyl groups (see Fig. 9.3). Analogous protein-residue side chains include alanine, valine, leucine, cysteine, methionine, and even larger nonpolar groups, many juxtaposed at the solvent interface. It seems reasonable to expect that lattice-ordered

HYDRATE POLYHEDRA

	Dodecahedron	Tetrakai decahedron	Pentakai decahedron	Hexakai decahedron
Faces	12	14	15	16
Vertices	20	24	26	28
Edges	30	36	39	42
Volume enclosed	160 Å^3	230 Å^3	260 Å^3	290 Å^3

Figure 9.3 A selection of types of water polyhedra, with enclosures varying in interval volume from 160 Å^3 to 1600 Å(18). Other known types can enclose even larger volumes (see next page). At each vertex of a polyhedron there is an oxygen atom. Between adjacent vertices, ie, along each edge, there is a hydrogen atom in an O—H···O bond. These clathrate hydrates form enclosures for guest molecules of a wide range of structures. The structure at the bottom (next page) is that for $(i\text{-}C_5H_{11})_4N^+F^- \cdot 38H_2O$; it shows the orientation of the tetraisoamylammonium cation within lobes of fused (two 14- and two 15-) polyhedra of water molecules.

MULTIPLE FUSED POLYHEDRA

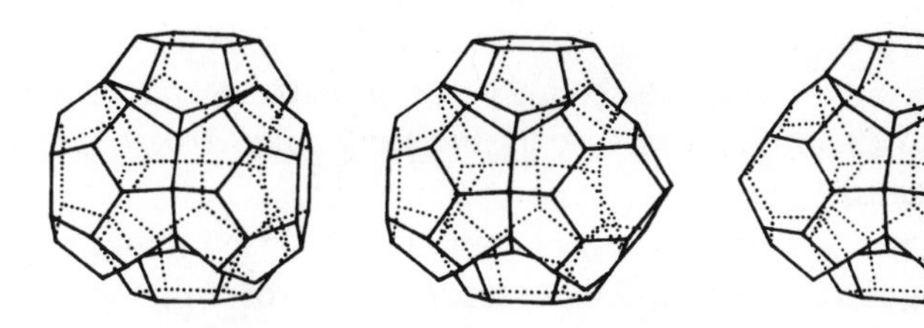

Fusion of	4(14-hedra)	3(14-hedra) 1(15-hedron)	2(14-hedra) 2(15-hedra)
Faces	44	45	46
Vertices	70	72	74
Edges	112	115	118
Volume enclosed	1000 Å^3	1000 Å^3	1000 Å^3

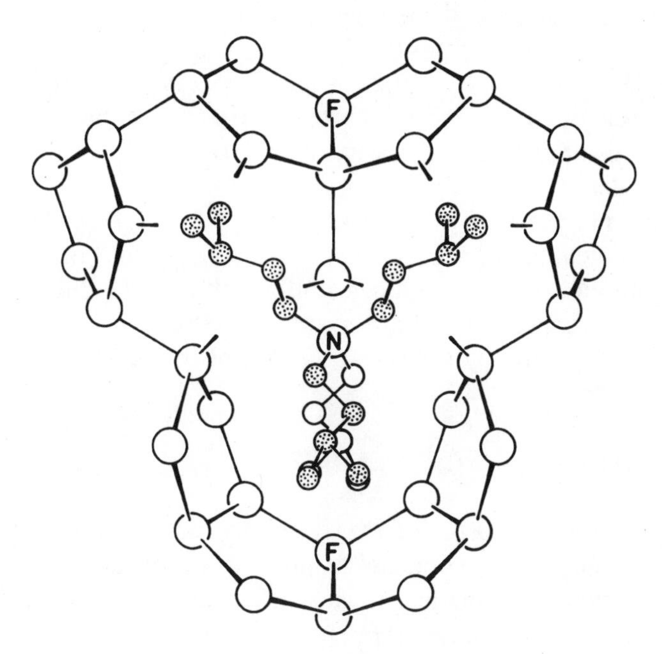

Figure 9.3 (*Continued*)

hydrate structures may be formed at the interface with the macromolecule (18). As Figure 9.3 illustrates, polyhedral surfaces in water clathrates can be quite varied in molecular geometry, clearly adapting to the structure of the enclosed entity. Thus it is apparent that water is a remarkably versatile substance in regard to hydrate formation, being capable of forming a large variety of cagelike structures to accommodate a whole gamut of solute molecules.

In recent years, resolutions in X-ray crystallography of proteins have improved remarkably, in some cases reaching below 1.0 Å; and ordered water molecules have been visualized (2). In crambin, a resolution near 0.9 Å has been approached and diffraction studies have disclosed an array of conjoint pentagonal water rings forming a clathrate hydrate structure in a region of the protein surface with a strong presence of nonpolar residues. Pentagonal rings of water have also been revealed in insulin and cytochrome *c*. It seems reasonable to expect, therefore, that as improved resolutions are achieved with other crystals, clathratelike water structures will also be revealed in other protein molecules.

The formation of polyhedral hydrates around nonpolar residues should contribute to the stabilization of a protein molecule and of its ligand complexes. That the energetics of hydrate formation are favorable thereto is evident from the fact that in the presence of appropriate simple guest molecules (see Table 9.4), crystal hydrates appear spontaneously and, depending on the enclosed molecule, are stable to temperatures above 30°C. Clearly, ΔG for hydrate formation is a negative number. The enthalpy changes observed for model compounds are in the range of -5 to -10 kcal mol^{-1} (18, 23) so the internal energy change favors formation. Since water molecules in the clathrate hydrate are more ordered and immobilized than in the bulk liquid, the entropy change presents a barrier, but in the model compounds it is obviously overcome by the favorable ΔH of the reaction.

9.5. INTEGRATING PERSPECTIVE

With this survey of forces of interaction as background, let us return to the assembly of experimental results shown in Table 8.2.

Starting with the simplest ligands, chloride and thiocyanate ions, we note that in both cases ΔH° is very small and ΔS° is positive. Since Cl^- and SCN^- are simple charged ions, it is hardly appropriate to label them as hydrophobic, yet the reaction is entropy driven. Similar conclusions follow from the data for cupric ion binding. Thus we see that complex formation involving charged species has thermodynamic parameters similar to those for binding of nonpolar hydrophobic groups. So a positive ΔS° for ligand binding is not diagnostic of hydrophobic interactions.

The thermodynamic data for the series of ligands CH_3COO^- to $CH_3(CH_2)_6COO^-$ reveal another interesting feature. With the shorter nonpolar substituents, CH_3- and $CH_3(CH_2)_4$-, ΔH° is negligible and the positive ΔS° drives the ligand–receptor complexation. With increasing length of the nonpolar, presumably hydrophobic substituent, ΔH° increases in (negative) numerical value and ΔS° drops. In other words, the change in internal energy becomes increasingly favorable and the contribution of ΔS° progressively less.

If we scan all the data in Table 9.2, listing increasingly larger ligands with many nonpolar substituents, we note that for many of the complexes, the favorable ΔG° values of formation are associated with positive entropy changes. However, for many others ΔS° is a large negative value, and the complex would not have formed were it not for the dominant effect of a favorable (i.e., negative) ΔH°. Again, we see that ligands with nonpolar ligands, presumably hydrophobic, form complexes with receptors because the change in internal energy provides the driving force, not the entropy change.

It is essential to recognize that in a molecule as complex as a receptor, views of bonding derived from ideal, isolated, model structures are inadequate. Substantial energy has been expended during biosynthesis to bring the constituent residues together by covalent linkage. The primary structure then provides the foundation for local, regional, and global interactions. Furthermore, the receptor and ligand are immersed in an aqueous environment; the Gibbs-Duhem thermodynamic constraint requires that any perturbation of the chemical potential of the solvent will also be transmitted to the solute receptor and ligand (24). For example, hydration can affect torsional transitions in a protein (25).

Hydrogen bonding obviously occurs in proteins, but the intrinsic stability of such interactions in that environment is uncertain. There can be coupled perturbations from neighboring residues. Other more distant peptide groups with their large dipole moments must have some effect on hydrogen bonds in the biomacromolecule. Furthermore, amide groups may bond not only to each other but to solvent water molecules, and the differences in strengths of the hydrogen bonds are probably small. Entropy changes for the dissociation of a hydrogen bond in a simple model compound can hardly be representative of those for a hydrogen bond in a protein where configurational and conformational changes of the biomacromolecule are coupled with those in the residue on which one is focusing.

Similar ambiguities are encountered in estimating the significance of charge–charge interactions. In general, ionic groups of proteins are on the surface of the macromolecule. In general, also, small anions such as Cl^- or SCN^- will not form complexes in bulk water with cations such as Na^+ or NH_4^+. But Cl^- and SCN^- are bound by serum albumin. Phosphate ion is bound by hemoglobin; ClO_4^- and NO_3^-, by hemerythrin. It becomes abundantly evident that charge–charge interactions play a role in ligand–receptor binding. It is also clear that a protein surface or matrix is not like unperturbed solvent water. Attempts have often been made to assign an effective dielectric constant, which must be different at the protein surface or interior than in bulk solvent; but the appropriateness of such a continuum viewpoint is dubious. Charge–charge interactions in ligand–protein complexes can be much more stabilized than are corresponding ion pairs in bulk water because of the possible orientations of dipoles within the biomacromolecule (26). The very fact that the receptor provides a region that excludes or diminishes bulk water when the ligand is inserted will stabilize ionic bonds by interactions with the internal permanent dipoles in the protein.

The net effect of van der Waals interactions on ligand–receptor stability, like hydrogen bonding, depends on small differences in the interaction energy of each constituent with its pair partner as compared with bulk water. Although the net energy may be only a fraction (eg, −0.2 kcal) per atom mole, the larger the ligand, the more atoms are involved—from it and the receptor—in the overlap. Thus an appreciable stabilization energy may appear. In addition, it is possible that dipole–dipole interactions between ligand and receptor groups may be enhanced in the protein environment. There have also been indications that aromatic side chains of proteins can be involved in specific aromatic–aromatic interactions between pairs, or multiplets, of aromatic rings (27, 28). Furthermore, an aromatic ring can form hydrogen bonds with water and with N—H groups (3–6). Also one should not overlook the possibility that specific *repulsive* van der Waals interactions within the receptor may be removed when the ligand is bound.

Ambiguous also are the possible contributions to configurational entropy and dynamics of the protein molecule when ligand is bound. In addition, solvent effects may influence these perturbations (29).

Accessible macromolecular conformations and their flexibility, transitions, and dynamics are subjects of current theoretical explorations.

Even in this brief, qualitative survey it can be recognized that many components are involved in biomolecular bonding and that the net stability is the resultant of a large number of coupled weak interactions. It is imprudent to assert that any one type of interaction is dominant. Nevertheless, it is evident that the interplay of many competing and reinforcing weak bonds poises the big macromolecule complex in a state that is sensitive and responsive to small changes. It is apparent experimentally that a small alteration in the structure of a ligand (eg, lengthening of an apolar chain) can result in a large increase (or decrease) in the stability of the complex with the receptor. This feature is a thermodynamic consequence. A change of slightly more than -1 kcal in ΔG° will increase the stability (equilibrium) constant K by a factor of 10 ($\Delta G^\circ = -RT \ln K$). So it is possible to modulate a physiological response with a small variation in chemical stimulus. Appropriate physicochemical analysis of ligand–receptor binding can thus lay the foundation for a molecular probing of the cellular consequences of the coupling.

REFERENCES

1. J. A. V. Butler, *Trans. Faraday Soc.,* **33,** 229–238 (1937).
2. I. M. Klotz, *Protein Sci.,* **2,** 1992–1999 (1993).
3. J. Sunner, K. Nishizawa, and P. Kebarle, *J. Phys. Chem.,* **85,** 1814–1820 (1981).
4. D. A. Rodham, S. Suzuki, R. D. Suenram, F. J. Lovas, S. Dasgupta, W. A. Goddard III, and G. A. Blake, *Nature*, **362,** 735–737 (1993).
5. R. N. Pribble and T. S. Zwier, *Science,* **265,** 75–79 (1994).
6. D. A. Dougherty, *Science,* **271,** 163–168 (1996).
7. I. M. Klotz, *J. Phys. Chem.,* **55,** 101–111 (1951).
8. R. M. Fuoss, *J. Am. Chem. Soc.,* **58,** 982–984 (1936).
9. J. G. Kirkwood, *Chem. Rev.,* **19,** 275–307 (1936).
10. F. H. Westheimer and J. G. Kirkwood, *Trans. Faraday Soc.,* **43,** 77–87 (1947).
11. J. Wyman Jr., and E. N. Ingalls, *J. Am. Chem. Soc.,* **60,** 1182–1184 (1938).
12. W. M. Latimer and W. H. Rodebush, *J. Am. Chem. Soc.,* **42,** 1419–1433 (1920).

13. S. N. Vinogradov and R. H. Linnell, *Hydrogen Bonding,* Van Nostrand Reinhold Co., New York, 1971, pp. 121–123.
14. I. M. Klotz and J. S. Franzen, *J. Am. Chem. Soc.* **84,** 3461–3466 (1962).
15. J. S. Franzen and R. E. Stephens, *Biochemistry* **2,** 1321–1327 (1963).
16. F. Hibbert and J. Emsley, *Adv. Phys. Org. Chem.,* **26,** 255–379 (1990).
17. G. C. Kresheck and I. M. Klotz, *Biochemistry,* **8,** 8–12 (1969).
18. I. M. Klotz, *Brookhaven Symp. Biol.,* **13,** 25–43 (1960).
19. P. Y. S. Lam and co-workers, *Science,* **263,** 380–384 (1994).
20. D. Liu, J. D. Cruzan, and R. J. Saykally, *Science,* **271,** 929–933 (1996).
21. H. Davy, *Ann. Chem. (Paris),* **78,** 298 and **79,** 5–28 (1811).
22. M. Faraday *Quant. J. Sci.,* **15,** 71 (1823).
23. I. M. Klotz in G. E. W. Wolstenholme and M. O'Connor, eds. *Ciba Foundation Symposium on the Frozen Cell,* J. A. Churchill, London, 1970, pp. 5–26.
24. I. M. Klotz, *Arch. Biochem. Biophys.* **116,** 92–96 (1966).
25. P. J. Steinbach and B. R. Brooks, *Proc. Natl. Acad. Sci. USA,* **90,** 9135–9139 (1993).
26. A. Warshel, *Proc. Natl. Acad. Sci. USA,* **75,** 5250–5254 (1978).
27. S. K. Burley and G. A. Petsko, *Science,* **229,** 23–28 (1985).
28. S. K. Burley and G. A. Petsko, *FEBS Lett.,* **203,** 139–143 (1986).
29. B. J. Berne, *Proc. Natl. Acad. Sci. USA*, **93,** 8800–8803 (1996).

CHAPTER 10

MOLECULAR SCENARIOS

If molecular pictures are joined to energetic perspectives, insights are gained into possible structural reasons for the variation of binding affinities with concentration of ligand. This approach provides a framework for understanding physiological and cellular regulatory responses to different concentrations of an effector.

With a single-site receptor, the binding graphs B versus L or B versus log L follow a rectangular hyperbola and an S-shaped curve of ideal shape, respectively (see Figs. 2.5 and 2.6). Deviations from the ideal are observed only with multivalent receptors. Two different molecular scenarios, out of the many that might be envisaged, will be examined. The molecular principle underlying each of these pictures can be elaborated with even the simplest nonideal system, a divalent ligand–receptor complex for which the affinity changes after the first ligand is bound.

If the receptor has two identical binding sites with fixed, unchanging affinities, then the uptake of ligand will fit the ideal binding curves shown in Figures 2.5 and 2.6. In most real cases, however, the binding deviates from the ideal curves. For the most general, nonideal behavior, the site constants indexed in Figure 10.1 would all be different. Under these circumstances, the following relations between stoichiometric and site constants can be derived (see Appendix 3):

$$K_1 = k_1 + k_2 \tag{10.1}$$

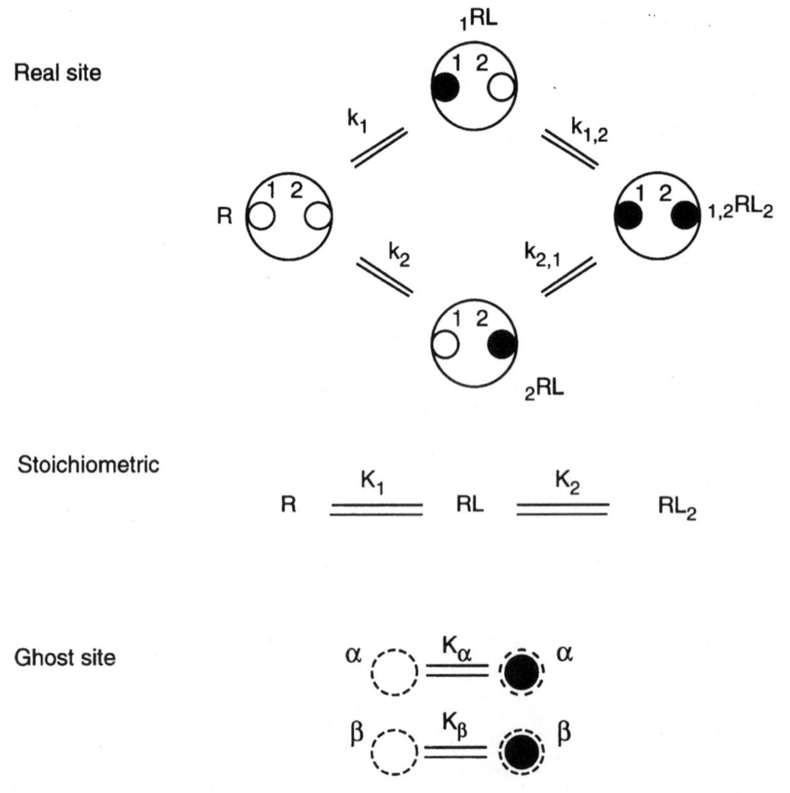

Figure 10.1 Schematic representation of divalent receptor with two binding sites in which each site affinity changes with binding of ligand at the other. Also shown are the stoichiometric and ghost-site binding constants.

$$K_1 K_2 = k_1 k_{1,2} \tag{10.2}$$

Since the three site constants are independent, their individual values cannot be fixed; only two stoichiometric constants are needed to specify the full course of a B versus L binding curve, and three unknowns cannot be determined from two constraints.

10.1. INITIALLY IDENTICAL SITES, AFFINITIES CHANGING WITH OCCUPANCY

Many receptors are constituted of two identical subunits. Under these circumstances (see Fig. 10.1),

$$k_1 = k_2 \equiv k_{\mathrm{I}} \tag{10.3}$$

Furthermore, as can be seen on inspection of Figure 10.1, the singly filled receptor species ${}_1RL$ and ${}_2RL$ must also be identical. Hence, insofar as the affinities for binding of the second ligand are concerned,

$$k_{1,2} = k_{2,1} \equiv k_{\mathrm{II}} \tag{10.4}$$

Nevertheless, unless we wish to impose another assumption, then we should recognize that in general

$$k_{\mathrm{I}} \neq k_{\mathrm{II}} \tag{10.5}$$

i.e., the (identical) first-step affinities need not be the same as the (identical) second-stage affinities.

The structural information that leads to equations (10.3) and (10.4) can now be inserted into equations 10.1 and 10.2. Thereby we can readily obtain the relations between the sequential stage affinities k_{I} and k_{II} and the stoichiometric binding constants:

$$K_1 = 2k_{\mathrm{I}} \tag{10.6}$$

$$K_2 = \tfrac{1}{2}k_{\mathrm{II}} \tag{10.7}$$

An example of this type of molecular scenario is provided by the divalent ligand–receptor complex Ca^{++}/calbindin. The stoichiometric binding constants, computed from binding data (Table 6.1), are

$$K_1 = 2.2 \times 10^8$$

$$K_2 = 3.7 \times 10^8$$

Using equation 10.6 and 10.7 we find

$$k_{\mathrm{I}} = 1.1 \times 10^8 = k_1 = k_2$$

$$k_{\mathrm{II}} = 7.4 \times 10^8 = k_{1,2} = k_{2,1}$$

It is immediately evident that after either of the first two sites is occupied by ligand, the affinity in the second sequential molecular step is markedly accentuated.

To complete the analysis of a divalent complex, let us turn to the general relations between the virtual, ghost-site constants and the real site constants (see Appendix A4):

$$K_\alpha = \tfrac{1}{2}(k_1 + k_2) + \tfrac{1}{2}[(k_1 + k_2)^2 - 4(k_1 k_{1,2})]^{1/2} \qquad (10.8)$$

$$K_\beta = \tfrac{1}{2}(k_1 + k_2) - \tfrac{1}{2}[(k_1 + k_2)^2 - 4(k_1 k_{1,2})]^{1/2} \qquad (10.9)$$

The equation for moles bound ligand B in terms of virtual constants is (equation 4.7),

$$B = \frac{K_\alpha L}{1 + K_\alpha L} + \frac{K_\beta L}{1 + K_\beta L} \qquad (10.10)$$

Looking at equations 10.8 and 10.9, we see that even though equation 10.10 resembles that for a divalent receptor with two different but fixed affinities, the virtual binding constants K_α and K_β are certainly not assignable to individual site binding constants, k_1 or k_2 or $k_{1,2}$ or $k_{2,1}$.

In fact, for a divalent receptor with accentuating affinities as occupancy rises, K_α and K_β are complex numbers. For Ca^{++}/calbindin (see Table 6.1), e.g.,

$$K_\alpha = 2.9 \times 10^8 e^{0.38\pi i}$$
$$K_\beta \doteq 2.9 \times 10^8 e^{-0.38\pi i}$$

Even if equations 10.3 and 10.4 are appropriate, one still obtains imaginary numbers for K_α and K_β from equations 10.8 and 10.9, for

$$k_{1,2} = k_{\mathrm{II}} = 7.4 \times 10^8 > 1.1 \times 10^8 = k_{\mathrm{I}} = k_1 = k_2$$

It is also possible to have a divalent receptor with two identical unoccupied sites for which equations 10.3 and 10.4 are applicable but in which the binding affinity of the second sequential ligand uptake is weaker than that for the first, i.e.,

$$k_{\mathrm{II}} < k_{\mathrm{I}} \qquad (10.11)$$

Such receptors are often designated as systems with half-of-sites reactivity, i.e., although two identical binding sites are initially open to ligand, once the first substrate molecule is bound, the residual open site no longer is receptive toward a second ligand.

10.2. CONFORMATIONAL EQUILIBRIA: ALLOSTERIC INTERACTIONS

Another possible molecular scenario that can explain cooperativity is diagrammed in Figure 10.2 (1). The core postulate in the picture is that there exist two different conformational forms, *R* and *T*, of the

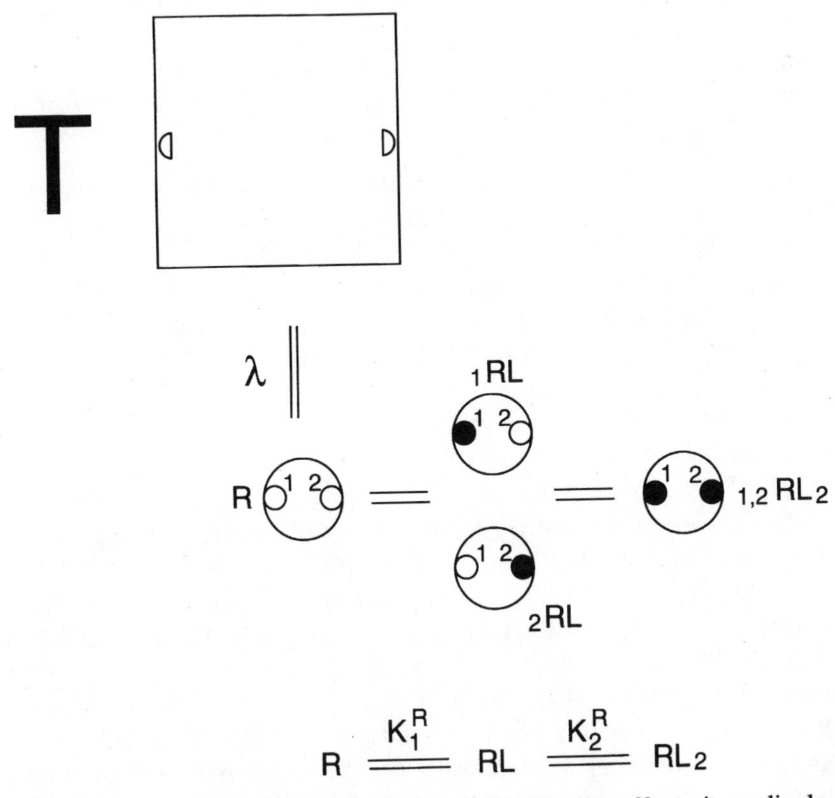

Figure 10.2 Conformation equilibria and allosteric effects in a divalent ligand–receptor system. Open sites on T form, ◁; open sites on R form ○; occupied sites on R form (●). Since the T form is assumed to be present in much larger concentration than the R form, the letter for T has been drawn as much larger than that for R.

receptor in equilibrium with each other:

$$R = T \tag{10.12}$$

Each can bind ligand *L*, but the relaxed *R* form is assumed to have a greater affinity than does the taut *T* form. This difference is represented schematically in Figure 10.2. The ligand is of a size that fits into the unoccupied sites in *R* but that does not lodge well in those provided by *T*.

To achieve the binding behavior we wish to explain, we also assume that the concentration of the *T* form is much larger than that of *R*. For subsequent mathematical manipulations, we define the conformational equilibrium constant λ as

$$\lambda = \frac{(T)}{(R)} \tag{10.13}$$

and keep in mind that for noticeable allosteric effects λ must be a large number. Keeping Figure 10.2 in sight, we can present a qualitative exposition of the essence of the allosteric influence on sequential affinities of receptor for ligand.

If conformation *T* did not exist, then when some ligand *L* is bound to empty sites on *R*, the equilibrium represented by K^R, is shifted to the right and the concentration of empty site *R* is reduced. Consequently, when more *L* is introduced, the number of open sites available to it is fewer and the probability of its being bound is reduced. So in a graph of *B* versus *L*, the rise will be diminished as more *L* is added; and in an ideal (divalent) system with identical, unchanging sites, a hyperbolic binding curve will be generated.

On the other hand, if another conformation *T* of the receptor is also present in equilibrium with *R*, then as ligand is bound to empty *R* to produce *RL*, the depleted *R* will be regenerated by a shift to the right in the $T = R$ equilibrium. If *T* is in large excess it will essentially fully resupply *R* without in itself being appreciably depleted. Consequently, the original concentration of open sites will be reproduced. Furthermore, in a divalent system some ${}_1RL$ and ${}_2RL$ will have been generated, and each of these supplies additional empty binding sites. Thus the net result of the addition of the first portion of *L* is to generate more open and available *R* binding sites accessible to the second portion of *L* than existed before the initial ligand was presented.

Consequently, more binding of the second portion takes place than occurred with the first injection. The binding graph curves upward in comparison to the ideal curve. Such accentuated binding is often described anthropomorphically as cooperativity.

This allosteric scenario can also be elaborated on in an analytical, quantitative format. Following the procedure used in Chapter 3, we write the stoichiometric equilibrium constants for binding of ligand L by the R and T forms, respectively:

$$R + L = RL; K_1^R = \frac{(RL)}{(R)(L)}; (RL) = K_1^R(R)(L) \tag{10.14}$$

$$RL + L = RL_2; K_2^R = \frac{(RL_2)}{(RL)(L)}; (RL_2) = K_1^R K_2^R(R)(L)^2 \tag{10.15}$$

$$T + L = TL_1; K_1^T = \frac{(TL)}{(T)(L)}; (TL) = K_1^T(T)(L) \tag{10.16}$$

$$TL + L = TL_2; K_2^T = \frac{(TL_2)}{(TL)(L)}; (TL_2) = K_1^T K_2^T(T)(L)^2 \tag{10.17}$$

The stage is now set for expressing the moles of bound ligand in terms of the contributions thereto by the various liganded species:

$$B = \frac{(RL + 2RL_2) + (TL + 2TL_2)}{(R + RL + RL_2) + (T + TL + TL_2)} \tag{10.18}$$

With appropriate substitutions from equations 10.14 to 10.17, we obtain

$$B = \frac{R(K_1^R L + 2K_1^R K_2^R L^2) + T(K_1^T L + 2K_1^T K_2^T L^2)}{R(1 + K_1^R L + K_1^R K_2^R L^2) + T(1 + K_1^T L + K_1^T K_2^T L^2)} \tag{10.19}$$

Since within each conformation the binding sites are identical and nonchanging in affinity, the successive stoichiometric constants are statistically related to an intrinsic constant K^R or K^T, respectively:

$$K_1^R = 2K^R; K_2^R = \tfrac{1}{2}K^R \tag{10.20}$$

$$K_1^T = 2K^T; K_2^T = \tfrac{1}{2}K^T \tag{10.21}$$

For convenience we also make the following substitution for the T factor in the numerator and denominator of equation 10.19:

$$T = \frac{R}{R} T = R\lambda \tag{10.22}$$

These specifications permit us to factor out R in the numerator and in the denominator, and then to write

$$B = \frac{[2K^R L + 2(K^R L)^2] + \lambda[2K^T L + 2(K^T L)^2]}{[1 + 2K^R L + (K^R L)^2] + \lambda[1 + 2K^T L + (K^T L)^2]} \tag{10.23}$$

$$B = \frac{2K^R L[1 + (K^R L)] + \lambda 2K^T L[1 + (K^T L)]}{[1 + (K^R L)]^2 + \lambda[1 + (K^T L)]^2} \tag{10.24}$$

Let us now examine some limiting special cases of the general equation 10.24.

1. Suppose both conformations R and T have the same affinity for ligand, i.e.,

$$K^R = K^T \equiv K \tag{10.25}$$

In that case, equation 10.24 becomes

$$B = \frac{2KL[1 + KL](1 + \lambda)}{[1 + KL]^2(1 + \lambda)} \tag{10.26}$$

$$B = \frac{2KL}{1 + KL} \tag{10.27}$$

This is the ideal binding equation for a divalent receptor with two identical sites of unchanging affinity (see equation 2.14). In other words, if the conformations R and T have the same affinity for ligand, they are indistinguishable in binding measurements, and the B versus L curve will be a rectangular hyperbola.

2. Suppose the T form has no affinity for ligand, i.e., $K^T = 0$ (but λ is not zero). Then equation 10.24 becomes

$$B = \frac{2K^R L[1 + (K^R L)]}{[1 + (K^R L)]^2 + \lambda} \tag{10.28}$$

In a B versus L graph this equation gives an S-shaped form, the signature of a system with cooperative interactions between binding sites. The larger the value of λ, the stronger the deviation of the binding curve from that for an ideal system.

3. Suppose that $\lambda = 0$. In that case, equation 10.24 again becomes that for a rectangular hyperbola (equation 10.27). That is hardly surprising since if $\lambda = 0$ then there is only one conformational state of the receptor, and it has two identical binding sites of nonchanging affinity.

Historically, the quantitative approach to allosteric effects was first formulated for the oxygenation equilibria of hemoglobin (1). The principles of the approach are similar to those illustrated in equations 10.12 to 10.24 for a dimeric system. Mammalian hemoglobin, however, occurs usually as a tetrameric receptor. Nevertheless, one can still postulate just two conformational states, R and T, so only one allosteric equilibrium constant λ (equation 10.13) is required. On the other hand, since four molecules of O_2 can be bound, one must write two sets of four stoichiometric steps for ligand binding in place of the two sets shown in equations 10.14 to 10.17. The algebra leading to an equation for B in terms of L has additional terms but is formally similar to that which led to equation 10.24. The result can be reduced to:

$$B = 4\,\frac{K^R L[1 + K^R L]^3 + \lambda K^T L[1 + K^T L]^3}{[1 + K^R L]^4 + \lambda[1 + K^T L]^4} \tag{10.29}$$

The graphical shapes of the curves generated from equation 10.29 depend on the values of the allosteric constant λ, and the relative affinities K^T/K^R of the T and R forms for O_2. As with the divalent receptor described above, it is illuminating to examine special cases.

For example, if both conformational forms have the same affinity for dioxygen,

$$K_T/K_R = 1 \tag{10.30}$$

then equation 10.29 can be reduced to

$$B = \frac{4KL}{1 + KL} \tag{10.31}$$

the ideal binding equation for a tetravalent receptor with identical sites of nonchanging affinity. A graph of this binding curve is a rectangular hyperbola, as is shown in Figure 10.3 (top). As K^T becomes increasingly weaker than K^R, the deviation of the binding curve from a hyperbola increases and tends toward an S-shape. When K^T approaches zero, a classic S-shaped curve is evident.

Figure 10.3 (bottom) illustrates the increasing deviation from ideal hyperbolic curvature as the T form becomes increasingly dominant in the allosteric equilibrium. Particularly when K_T approaches zero, the taut conformation serves simply as a reservoir to replenish the R form, with its strong affinity for ligand, as complexes RL_1 to RL_4 are formed progressively.

The concept of conformational equilibria in protein structures appeared in qualitative explanations of protein behavior even during the first half of this century. The quantitative formulation of allosteric interactions (1), however, was formulated only approximately contemporaneously with the appearance of the first X-ray diffraction solutions of the three-dimensional structure of a protein. Thus it became possible promptly to search for different conformations of a particular protein. X-ray crystallography rapidly disclosed two major conformations of hemoglobin, corresponding to the R and T states of the allosteric analysis. Subsequently, diffraction studies revealed other structural forms (2). Electromagnetic spectroscopic techniques have also been developed to provide structural probes, and additional conformations have been discovered that are intermediates between the fully R and fully T forms of hemoglobin (3,4). Furthermore, alternative conformational forms have been found in many of the several hundred proteins whose three-dimensional structures have been fully resolved (5,6). Thus structural biology is reaching a stage where it can be coupled with the principles of ligand–receptor energetics to provide a firm conceptual bridge leading to detailed molecular understanding of regulation and control in cellular biology.

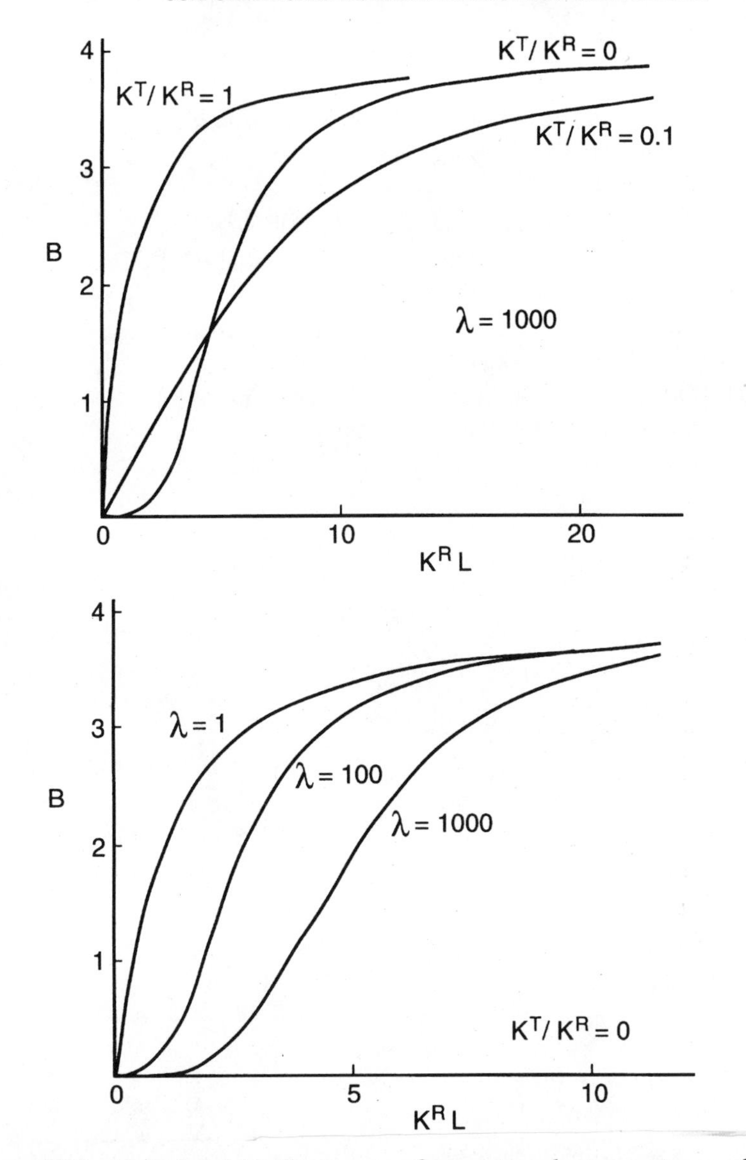

Figure 10.3 Calculated binding curves for a tetravalent receptor, such as hemoglobin $|O_2$, if it has two conformational forms T and R in an allosteric equilibrium. Top, the dependence of the shapes of curves on the relative affinities K^T and K^R for ligand L; the allosteric constant is taken as 1000, ie, overwhelmingly favoring the taut conformation. For the special circumstance of $K_T = K_R$, a rectangular hyperbola is seen. Bottom, the dependence of the shapes of curves on the magnitude of allosteric constant λ; the ligand affinity of the taut form T is taken as zero. In the limit of $\lambda = 0$, a rectangular hyperbola is seen.

REFERENCES

1. J. Monod, J. Wyman, and J.-P. Changeux, *J. Mol. Biol.* **12,** 88–118 (1965).
2. M. M. Silva, P. H. Roger, and A. Arnone, *J. Biol. Chem.,* **267,** 17248–17256 (1992).
3. C. Ho, *Adv. Prot. Chem.,* **43,** 153–312 (1992).
4. V. Jayaraman, K. R. Rodgers, I. Mukerji, and T. G. Spiro, *Science,* **269,** 1843–1848 (1995).
5. C. Branden and J. Tooze, *Introduction to Protein Structure,* Garland Publishing Co., New York, 1991.
6. D. Huang, C. F. Ainsworth, F. J. Stevens, and M. Schiffer, *Proc. Natl. Acad. Sci. USA,* **93,** 7017–7021 (1996).

APPENDIX A1

EXPERIMENTAL PROCEDURES IN EQUILIBRIUM DIALYSIS

A1.1. COMPONENTS OF APPARATUS

A1.1.1. Semipermeable Membrane

The membrane must be freely permeable to water and to small molecules and ions but at the same time must not permit passage of the receptor macromolecule (1,2). Four sources for appropriate cellulose tubing membranes are the Union Carbide Corp., Spectrum Medical Industries, Inc., Sartorius GmbH, and Pierce Chemical Co. A range of sizes is available. That of $\frac{8}{32}$ inch (inflated diameter) is convenient for a volume of 1 mL for the macromolecule solution. For 5 mL volumes, $\frac{18}{32}$ inch is suitable; for 10 to 20 mL samples, $\frac{23}{32}$ inch is generally used. For flat membranes, cut out pieces from large tubing, e.g., $\frac{36}{32}$ inch, in the geometric shape appropriate for the dialysis chambers. Some suppliers specify the molecular weight cutoff for permeability (generally 10,000 to 30,000). If there is any uncertainty in regard to a specific macromolecule, place a portion of receptor solution in one compartment of the dialysis apparatus and an equal volume of receptor-free solution in the other; test the latter (for 24 h) for any leakage of macromolecules into it.

A1.1.2. Containers

Pyrex test tubes of appropriate size serve as dialysis vessels when dialysis bags are used. For volumes of 5 to 20 mL of receptor solution, tubes of 25 × 200 mm are convenient (Fig. A1.1A). Use rubber stoppers to close the tubes to prevent evaporation. They should be covered with thin sheets of polyethylene to minimize any risk of contamination of the solution by the rubber stopper. Alternatively, close the dialysis tube with a rubber serum cap. Attach a fiberglass thread (no. E-181) to the dialysis bag to facilitate stirring and to permit easy removal of the bag from the tube. Sever the desired length of thread from the spool by melting it in the flame of a small burner; cutting the threads leads to unravelling of the fibers.

Alternative suitable containers are glass vials with rubber-lined screw caps, sample vials with polyethylene caps, and glass tubes constructed from a pair of male and female standard taper joints. To deal with quantities of solution on the order of 1 mL, use a cell in which a flat sheet of cellulose membrane is held between two flat, cylin-

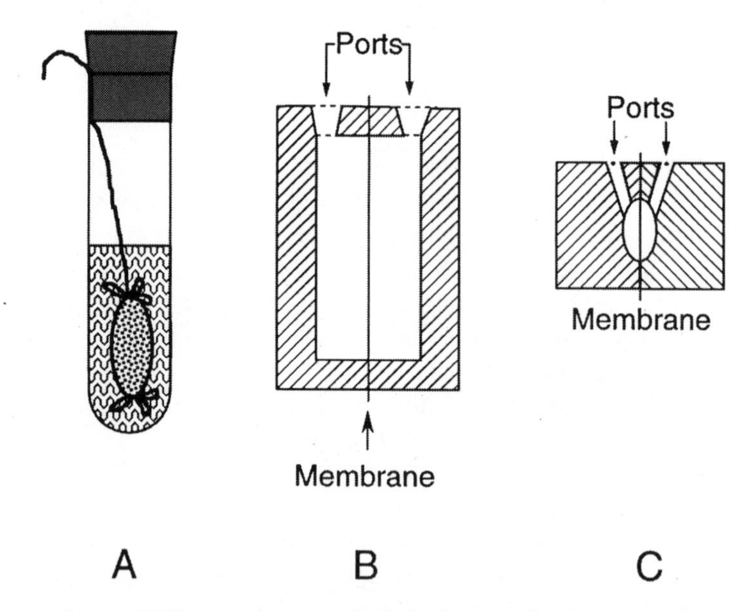

Figure A1.1 Different forms of dialysis vessels for measurement of ligand–receptor binding by equilibrium dialysis. **A,** A protein–containing dialysis bag is immersed in a protein-free solution contained in a test tube. This configuration is convenient for volumes of several or many milliliters. **B,** Two chambers are machined precisely to clamp a flat sheet of cellulose membrane between them. The volume of each chamber is about 1 mL. **C,** The principle of construction is similar to that in **B,** but the scale is much smaller and the volume in each chamber can be as small as 20 μL.

drical 1-mL compartments (3). The chambers can be fabricated from glass, metal, or plastics, such as Teflon, Plexiglas or polyacrylamide (Fig. A1.1B). Each cell compartment also has a port so that solution can be inserted or removed. Samples even smaller than 1 mL can be accommodated by a dialysis cell such as that shown in Figure A1.1C, which holds volumes of 20 μL in each chamber (4).

A1.1.3. Thermostat and Shaker

Adequate agitation is necessary to reach equilibrium in a matter of hours. Without agitation, the attainment of equilibrium may take days, especially with large volumes of sample. Efficient mixing is facilitated if an air bubble is enclosed within the dialysis bag or at the top of the sample compartments. All dialysis experiments should be carried out in a thermostat at a precise, fixed temperature. Although in many cases ligand binding does not strongly depend on temperature, any calculations of binding affinities and related thermodynamic quantities require specification of the precise temperature.

A1.2. OPERATIONS AND MANIPULATION

A1.2.1. Preparation of the Membrane

Cellulose tubing is generally manufactured by the xanthate process, which leaves substantial quantities of impurities in the product. These must be removed by extensive washing.

1. Take a roll of cellulose rubbing and cut off lengths appropriate to the volume of protein solution to be used in the dialysis experiment. For $\frac{23}{32}$-inch-diameter tubing, a 10-inch length is suitable for 10 to 20 mL of solution.
2. Place the cut tubing in a beaker of glass-distilled water and heat on a steam bath for 1 h.
3. Transfer the tubing to fresh water and repeat the heating. Then soak the tubing in fresh samples of water at room temperature for several hours; repeat this step twice.
4. Finally, transfer the tubing to a beaker containing the buffer solution or supporting electrolyte that will be used in the binding measurements and allow it to soak for 12 to 24 h. If the supporting electrolyte is a neutral salt instead of a buffer, adjust the

pH of the solution to the value desired for the dialysis experiment by adding dilute HCl; the tubing itself tends to lower the pH of an unbuffered solution. The tubing must be kept wet until it is used to form the dialysis bag.

Harsher treatment of the tubing than that described may impair the properties of the dialysis membrane. Heating tubing on a steam bath for 72 h weakens the bags and makes some permeable to macromolecular receptors of molecular weight as high as 70,000. Trace metal impurities may be diminished by extraction with chelators or dilute acid (pH 3).

A1.2.2. Solutions

A1.2.2.1. Donnan Effects If the receptor macromolecule carries a net charge and the ligand is also electrically charged, then, even in the absence of binding, the concentration of free ligand will not be the same in both compartments (Fig. A1.1) because of the Donnan effect. This difference in concentration can be made negligibly small by having an adequate quantity of buffer salt or neutral salt in the solution. For example, for the binding of a monovalent ion ligand by 0.05% serum albumin at pH 6.1, about 1 unit above its isoelectric point, 0.025 M phosphate buffer renders the Donnan effect negligible. At pH values farther from the isoelectric point, or at higher protein concentrations, higher concentrations of supporting electrolyte may be necessary.

A1.2.2.2. pH Control In general, a buffer must be used to maintain the pH at some defined value. For proteins with isoelectric points slightly below 7, a constant self-pH can be maintained, but electrolyte is still essential to swamp out the Donnan effect. Thus for serum albumin with an isoelectric point of 5, an appropriate neutral solvent is 0.1 M NaCl. The compositions of a few suitable buffers to establish pH values from 5 to 9 are listed in Table A1.1. An ionic strength of 0.1 will reduce the Donnan effect to a negligible value for most proteins at concentrations of 1% or less. Solutions of lower ionic strength at the same pH can be obtained by diluting the buffers shown in Table A1.1.

Prepare the solution of macromolecule in supporting electrolyte by dissolving crystalline or lyophilized material, if available. For a receptor with a ligand-binding equilibrium constant of 10^4 to 10^5 M^{-1}, its concentration should be about 0.2% by weight. The moisture content

TABLE A1.1 Composition of Some Buffers for Equilibrium Dialysis

Electrolyte	pH[a]	Molarity	Ionic Strength
Sodium acetate	5.0	0.100	0.100
Acetic acid		0.056	
Potassium acid phthalate	5.0	0.050	0.100
NaOH		0.025	
NaCl	5.0[a]	0.100	0.100
Citric acid	5.7	0.0234	0.100
NaOH		0.0567	
Mes	6.2	0.100	0.100
HCl		0.100	
Na_2HPO_4	6.8	0.035	0.132
KH_2PO_4		0.028	
Na_2HPO_4	6.9	0.0139	0.053
KH_2PO_4		0.0110	
Hepes	7.6	0.100	0.100
HCl		0.100	
Tris	8.3	0.100	0.100
HCl		0.100	
$NaHCO_3$	8.6	0.100	0.100
Glycine	9.1	0.400	0.100
NaOH		0.100	
Na_2CO_3	10.5	0.0345	0.100
HCl		0.0035	

[a]pH obtained by addition of 4.65 mL of 0.0025 M HCl to 500 mL of solution.

of the original solid macromolecule should be determined to calculate the actual concentration in solution. For this purpose, weigh a sample of about 10 mg into a small dish and place in a drying pistol *in vacuo* over a good desiccant, such as P_2O_5 at a temperature of 100°C until constant weight is reached. If some spectrophotometric or other assay for the receptor is well established, it can be used to determine the concentration. For example, if the molar absorption coefficient is known, an absorbance measurement provides directly the receptor concentration, provided that other absorbing substances are absent or can be compensated for.

Dissolve the pure ligand in the same supporting electrolyte as that for the receptor. For a ligand–receptor binding constant of 10^4 to 10^5 M^{-1}, the ligand concentration should cover a range of 10^{-4} to 10^{-5} M. Prepare a stock solution of 10^{-4} M and dilute aliquots with supporting electrolyte to obtain lower concentrations.

A1.2.2.3. Filling the Dialysis Tubes

1. Take a length of dialysis tubing from the final buffer solution in which it has been soaking. Remove adhering liquid promptly by placing the tubing between sheets of white absorbent paper and press out the liquid.
2. Make two tight knots at one end of the tubing.
3. Pipette in a precise quantity of protein solution (e.g., 10 mL), and force out most of the air above the liquid by manipulation with your fingers.
4. Close the top end of the dialysis bag by making two tight knots, the first of which should leave a bubble of air above the liquid.
5. Cut off the loose, open ends of the tubing close to the knots and discard these remnants.
6. Tie a fiberglass thread between the double knots at the top of the bag.
7. Pipette a precise quantity of ligand solution into the test tube (Fig. A1.1A).
8. Suspend the dialysis bag in the ligand solution with the free end of the thread hanging over the lip of the test tube; place the polyethylene-lined stopper securely in the tube, and mount the test tube in the shaker in the thermostat.

A1.2.2.4. Determination of Time for Equilibrium

1. Set up a series of five dialysis tubes, each containing the same concentration of receptor within the bag and the same concentration of ligand external to the bag.
2. Place each of the tubes in the thermostated shaker.
3. For the types of equipment illustrated in Figure A1.1, equilibrium is generally attained in 4 to 6 h. Therefore, remove one of the tubes after 3, 4, 6, 12, and 18 h, respectively.

4. Analyze the external solution for ligand concentration to determine the time required for equilibrium.

It may still be convenient, however, to plan experiments so that the assembly of the dialysis tubes is completed near the end of a working day and equilibration is carried out overnight.

A1.2.2.5. Confirmation of Impermeability of Membrane to Protein Even if the manufacturer specifies a molecular weight above which a macromolecule will not pass through the membrane, it is prudent to confirm that the membrane will retain the receptor in the chamber to which it was originally added.

1. Set up a series of three dialysis tubes each containing the same concentration of receptor within the bag and the same concentration of (ligand-free) buffer external to the bag.
2. Place each of the tubes in the thermostated shaker.
3. After 18 h, remove the tubes and analyze for receptor by some spectrophotometric or sensitive colorimetric assay.

Ideally, no detectable amount of receptor should be present. In practice, if less than 0.1% of that originally present in the bag has leaked through, the membrane is acceptable. This experiment is carried out in triplicate (at least) because occasionally a single sample of cellulose membrane can be defective.

A1.2.2.6. Analysis of Solutions at Equilibrium At the end of the equilibration period, remove the dialysis bags from the container, take out an aliquot of external solution with a pipette, and carry out an analysis for the ligand. In general, each ligand has a well-established specific mode of analysis for the concentration range being studied. By analyzing the external solution one avoids complications arising from the presence of receptor.

A1.3. COMPUTATION OF BOUND LIGAND

A1.3.1. Correction for Binding to Membrane

To illustrate these calculations, a set of actual experimental data has been assembled (Table A1.2). Six protein-free containers were used

TABLE A1.2 Analyses and Computations for Binding of 2-azo-*p*-dimethylaniline by Bovine Serum Albumin

	Protein-Free Tubes					
	1_0	2_0	3_0	4_0	5_0	6_0
$(L)_0$, initial concentration of ligand in external compartment, M × 10^5	6.474	4.849	3.249	2.782	2.161	1.530
L_T total moles of ligand added, $\times 10^7$ [a]	6.474	4.849	3.249	2.782	2.161	1.530
(L'), concentration of free ligand at equilibrium, M $\times 10^5$	2.543	1.827	1.193	1.010	0.801	0.514
$V_T(L')$, total moles of free ligand in both compartments, $\times 10^7$ [a]	5.086	3.654	2.386	2.020	1.602	1.028
L'_M, moles of ligand bound by membrane, $\times 10^7$	1.388	1.195	0.863	0.762	0.559	0.502
	Protein-Containing Tubes					
	1	2	3	4	5	6
$(L)_0$, initial concentration of ligand in external compartment, M × 10^5	19.499	13.892	9.725	6.474	4.849	3.249
L_T, total moles ligand, $\times 10^7$ [a]	19.449	13.892	9.725	6.474	4.849	3.249
R_T, total moles receptor, $\times 10^7$	4.425	4.425	4.425	4.425	4.425	4.425
(L), concentration of free ligand at equilibrium, M $\times 10^5$	2.275	1.655	1.901	0.719	0.524	0.322
$V_T(L)$, total moles of free ligand in both compartments, $\times 10^7$ [a]	4.550	3.310	2.182	1.438	1.048	0.644
$L_T - V_T(L)$, total moles ligand bound to receptor and membrane, $\times 10^7$	14.899	10.582	7.543	5.036	3.801	2.605
L_M, moles ligand bound to membrane $\times 10^7$	1.382	1.120	0.847	0.621	0.491	0.310
L_R, moles ligand bound to protein, $\times 10^7$	13.517	9.462	6.696	4.415	3.310	2.295
B, moles bound ligand per mole of receptor	3.055	2.138	1.513	0.998	0.748	0.519

[a] The total volume V_T in each container was 20 mL, 10 mL in the dialysis bag and 10 mL in the external compartment.

to obtain data for membrane binding. Let us examine the numerical entries for container 1. A precise quantity (10 mL) of solution of ligand of known concentration $(L)_o$ (6.474×10^{-5} M), was added to the test tube container to constitute the external compartment. Thus the total moles of ligand added L_T must be $0.01 \text{ L} \times 6.474 \times 10^{-5} \text{ M} = 6.474 \times 10^{-7}$ mol. Then the dialysis bag containing a precise quantity (10 mL) of protein-free buffer was placed in the ligand solution in the test tube, and the container was shaken gently in a thermostat until equilibrium was reached. The bag was removed, and the ligand concentration (L'), was determined, in this case by measurement of the optical absorbance, and found to be 2.543×10^{-5} M. Therefore, the total moles of free ligand $V_T(L')$ in the entire test tube, dialysis bag plus external compartment, must be $(0.01 + 0.01)$ L $\times$ 2.543×10^{-5} M $= 5.086 \times 10^{-7}$ mol. Hence the amount bound by the membrane and other components of the container L'_M is, following equation 1.4, $(6.474 - 5.086) \times 10^{-7}$ or 1.388×10^{-7} mol. Thus this experiment established that at a free ligand concentration of 2.543×10^{-5} M, the correction for membrane binding in this specific type of dialysis apparatus is 1.388×10^{-7} mol.

Similarly, test tube 2_o established that at $(L') = 1.827 \times 10^{-5}$ M, $L'_M = 1.195 \times 10^{-7}$ mol, and so on for tubes 3_o to 6_o. From these numbers a graph is prepared of the membrane-binding correction L'_M as a function of free, unbound ligand concentration (L') (Fig. A1.2).

A1.3.2. Calculation of Ligand Bound to Receptor

Protein-containing solution was added to the dialysis bags, and the data are shown in Table A1.2. The initial loading concentrations of

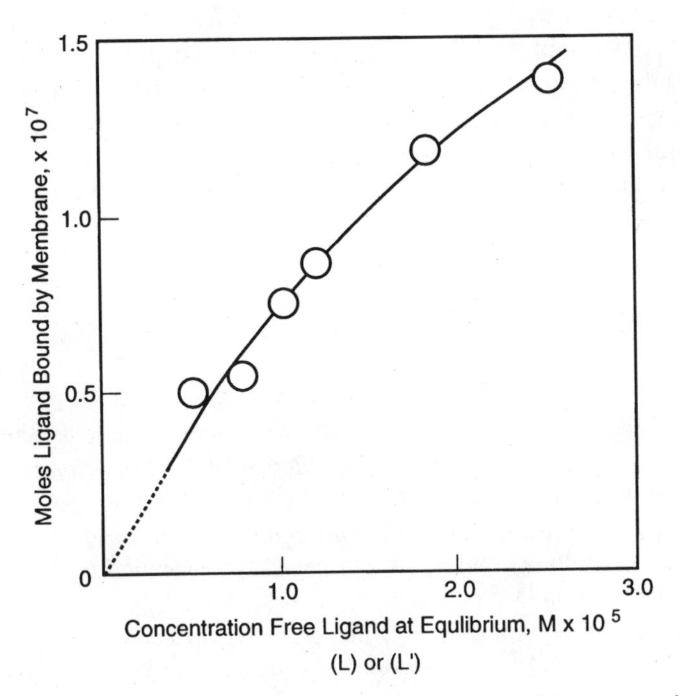

Figure A1.2 Example of a calibration curve for making corrections for binding of ligand to membrane and container.

ligand were higher than in the protein-free tubes, since a large fraction of the ligand would be bound by the receptor protein. Thus for test tube 1, 19.449×10^{-5} M was the initial ligand concentration in the 10 mL of protein-free solution added to the test tube, to constitute the external compartment in the dialysis apparatus. In this tube, therefore, the total moles of ligand L_T is 0.01 L $\times$ 19.449×10^{-5} M = 19.449×10^{-7} mol. The total receptor protein R_T added to the dialysis bag is 4.425×10^{-7} mol, and is the same for each tube because the same quantity of protein was used in each experiment. After equilibration, the concentration of free ligand (L) was determined by analysis of the external compartment. The same value must hold for both compartments at equilibrium. Therefore, the total moles of unbound, free ligand in tube 1 must be (0.01 + 0.01) L $\times$ 2.275×10^{-5} M mol. Hence the ligand removed from free solution by binding to the receptor and to the membrane, $L_T - V_T(L)$ is $(19.449 - 4.550) \times 10^{-7} = 14.899 \times 10^{-7}$ mol. The correction for binding by the membrane L_M is now found (see equation 1.4). That quantity is a function of the free ligand concentration (L), which in tube 1 is 2.275×10^{-5} M. At this concentration, according to Figure A1.2, L_M is 1.382×10^{-7} mol, which is then subtracted from 14.899×10^{-7} mol to give 13.517×10^{-7} mol, the ligand bound by the receptor L_R. Finally, to obtain the extent of binding B, divide 13.517×10^{-7} mol by 4.425×10^{-7} mol, as specified by equation 1.1. The extent of binding B is 3.055 mol ligand/mol receptor for the experiment in test tube 1. The corresponding sequences of computations for tubes 2 to 6 are shown in Table A1.2.

REFERENCES

1. I. M. Klotz, in *Protein Function: A Practical Approach*, T. E. Creighton, ed., IRL Press, Oxford, UK, 1989, pp. 25–54 and references therein.
2. D. J. Winzor and W. H. Sawyer, *Quantitative Characterization of Ligand Binding*, Wiley-Liss, New York, 1995.
3. F. Karush and S. S. Karush in *Methods in Immunology and Immunochemistry*, C. A. Williams and M. W. Chase, eds., Academic Press, Inc. New York, 1971, pp. 383–394.
4. R. Brodersen, S. Andersen, C. Jacobsen, O. Sønderskov, F. Ebbesen, W. J. Cashore, and S. Larsen, *Anal. Biochem.*, **121,** 395–405 (1982).

APPENDIX A2

PROPERTIES OF GRAPHICAL REPRESENTATIONS OF MULTIPLE CLASSES OF INDEPENDENT BINDING SITES

For a single class of identical, invariant binding sites, the following equations are applicable (see Chapter 2):

$$B = \frac{nkL}{1 + kL} \tag{A2.1}$$

$$\frac{1}{B} = \frac{1}{n} + \frac{1}{nk}\frac{1}{L} \tag{A2.2}$$

$$\frac{B}{L} = kn - kB \tag{A2.3}$$

The linear graphical equivalents of equations A2.2 and A2.3 are shown in Figure 2.7.

When there are two or more classes of identical, invariant sites, graphs of B/L versus B and of $1/B$ versus $1/L$ are no longer linear. Nevertheless, it is still possible to determine limiting slopes and intercepts for the experimental curves (1). However, the relations of these parameters to the number and to the binding constants of the individual classes of sites are not what is commonly assumed on intuitive grounds.

Let us consider first two classes of identical invariant sites. The intercepts and slopes that we wish to establish are shown in Figures A2.1 and A2.2.

To analyze the curve in Figure A2.1 for the case where $n = 2$, we write the algebraic form for two classes of sites,

$$B = \frac{n_1 k_1 L}{1 + k_1 L} + \frac{n_2 k_2 L}{1 + k_2 L} \tag{A2.4}$$

which can be converted into

$$\frac{B}{L} = \frac{n_1 k_1}{1 + k_1 L} + \frac{n_2 k_2}{1 + k_2 L} \tag{A2.5}$$

From the latter it follows that

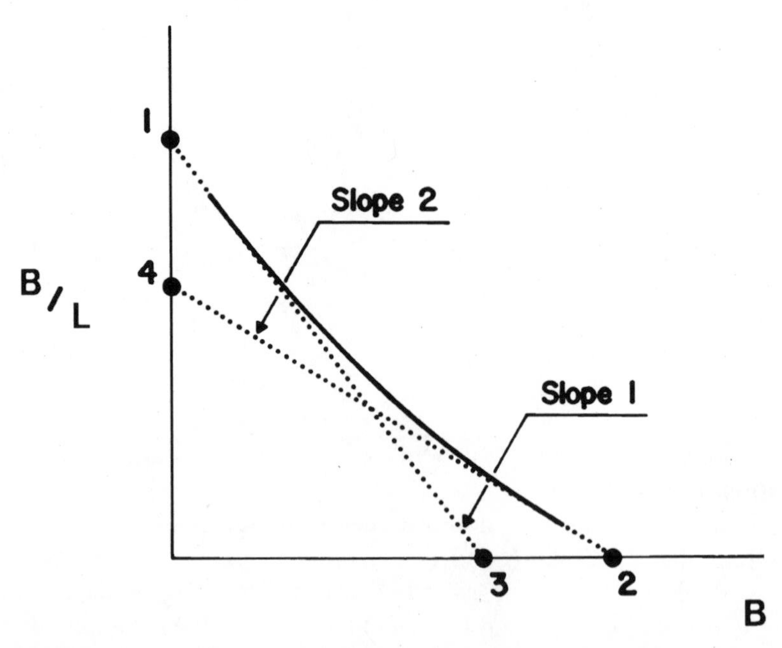

Figure A2.1 A schematic curve with intercepts and limiting slopes for a graph of variables B/L and B, where m independent classes of sites are present.

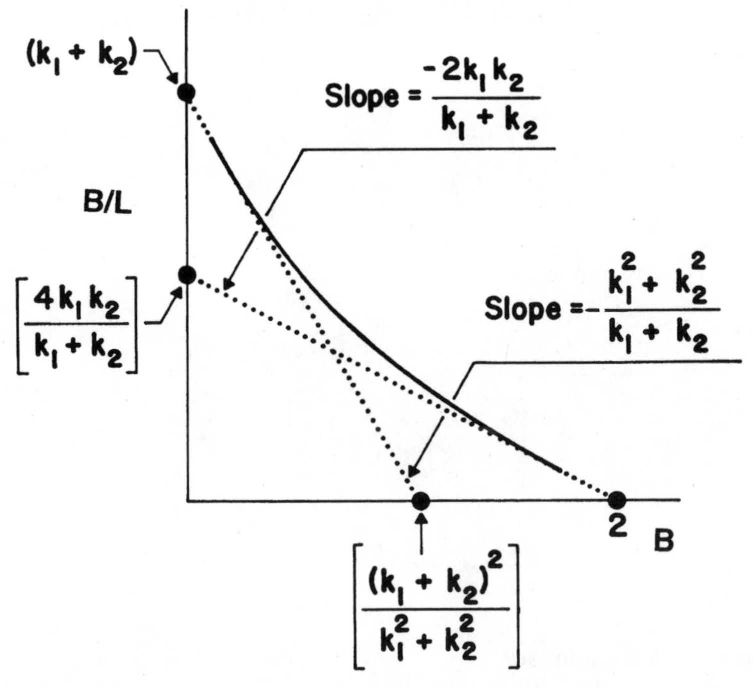

Figure A2.2 Schematic curve for a system of only two independent binding sites. The intercepts and limiting slopes are indicated.

$$\text{intercept } 1 = \lim_{\substack{B \to 0 \\ L \to 0}} \left(\frac{B}{L}\right) = n_1 k_1 + n_2 k_2$$

$$= n_T \frac{n_1 k_1 + n_2 k_2}{n_T} \qquad \text{(A2.6)}$$

where n_T is the total number of sites, the sum of the number in each of the two classes:

$$n_T = n_1 + n_2 \qquad \text{(A2.7)}$$

Consequently,

$$\text{intercept } 1 = n_T \langle k \rangle_1 \qquad \text{(A2.8)}$$

where $\langle k \rangle_1$ defines an average site binding constant for two classes of identical, invariant sites,

$$\langle k \rangle_1 = \frac{\sum_1^2 n_j k_j}{\sum_1^2 n_j} \tag{A2.9}$$

Approaching intercept 2 we write

$$B = \frac{n_1 k_1 L}{1 + k_1 L} \frac{\frac{1}{L}}{\frac{1}{L}} + \frac{n_2 k_2 L}{1 + k_2 L} \frac{\frac{1}{L}}{\frac{1}{L}} \tag{A2.10}$$

$$= \frac{n_1 k_1}{\frac{1}{L} + k_1} + \frac{n_2 k_2}{\frac{1}{L} + k_2} \tag{A2.11}$$

As $B/L \to 0, L \to \infty$; so

$$\text{intercept } 2 = \lim_{L \to \infty} B = \frac{n_1 k_1}{k_1} + \frac{n_2 k_2}{k_2} = n_T \tag{A2.12}$$

Turning to the limiting slope as $B \to 0$, i.e., slope 1, we start by recognizing that in general

$$\frac{d(B/L)}{dB} = \frac{\frac{d(B/L)}{dL}}{\frac{dB}{dL}} \tag{A2.13}$$

For the numerator of equation (A2.13), we obtain from equation (A2.5)

$$\frac{d(B/L)}{dL} = -\frac{n_1 k_1}{(1 + k_1 L)^2} k_1 - \frac{n_2 k_2}{(1 + k_2 L)^2} k_2$$

$$= -\frac{n_1 k_1^2}{(1 + k_1 L)^2} - \frac{n_2 k_2^2}{(1 + k_2 L)^2} \tag{A2.14}$$

For the denominator of equation A2.13 we start with equation A2.4:

$$\frac{dB}{dL} = \frac{n_1k_1}{1 + k_1L} - \frac{n_1k_1Lk_1}{(1 + k_1L)^2} + \frac{n_2k_2}{1 + k_2L} - \frac{n_2k_2Lk_2}{(1 + k_2L)^2} \tag{A2.15}$$

This can be reduced to

$$\frac{dB}{dL} = \frac{n_1k_1}{(1 + k_1L)^2} + \frac{n_2k_2}{(1 + k_2L)^2} \tag{A2.16}$$

Returning to equation A2.13 we write

$$\frac{d(B/L)}{dB} = -\frac{\dfrac{n_1k_1^2}{(1 + k_1L)^2} + \dfrac{n_2k_2^2}{(1 + k_2L)^2}}{\dfrac{n_1k_1}{(1 + k_1L)^2} + \dfrac{n_2k_2}{(1 + k_2L)^2}} \tag{A2.17}$$

This equation can be reduced further by straightforward algebraic manipulation to give the limiting slope as $B \rightarrow 0$, which is slope 1 in Figure A2.1,

$$\text{slope } 1 = \lim_{\substack{B \to 0 \\ L \to 0}} \left[\frac{d(B/L)}{dB}\right] = -\frac{n_1k_1^2 + n_2k_2^2}{n_1k_1 + n_2k_2} = -\langle k \rangle_2 \tag{A2.18}$$

where $\langle k \rangle_2$ defines a second type of average site binding constant.

To approach slope 2, we write the following equation as a further modification of equation A2.17:

$$\frac{d(B/L)}{dB} = -\frac{\dfrac{n_1k_1^2}{(1 + k_1L)^2}\dfrac{1}{L^{-2}} + \dfrac{n_2k_2^2}{(1 + k_2L)^2}\dfrac{1}{L^{-2}}}{\dfrac{n_1k_1}{(1 + k_1L)^2}\dfrac{1}{L^{-2}} + \dfrac{n_2k_2}{(1 + k_2L)^2}\dfrac{1}{L^{-2}}} \tag{A2.19}$$

$$= -\frac{\dfrac{n_1k_1^2}{(1/L + k_1)^2} + \dfrac{n_2k_2^2}{(1/L + k_2)^2}}{\dfrac{n_1k_1}{(1/L + k_1)^2} + \dfrac{n_2k_2}{(1/L + k_2)^2}} \tag{A2.20}$$

Now we are prepared to find the limiting slope as $L \rightarrow \infty$, i.e., as $B \rightarrow$ intercept 2 ($= n_T$):

$$\text{slope 2} = \lim_{\substack{L \rightarrow \infty \\ B \rightarrow n_T}} \frac{d(B/L)}{dB} = -\frac{n_1 + n_2}{\dfrac{n_1}{k_1} + \dfrac{n_2}{k_2}} = -\langle k \rangle_0 \tag{A2.21}$$

where $\langle k \rangle_0$ denotes a third type of average site binding constant,

$$\langle k \rangle_0 = \frac{n_1 + n_2}{\dfrac{n_1}{k_1} + \dfrac{n_2}{k_2}} \tag{A2.22}$$

With intercepts 1 and 2 and slopes 1 and 2 evaluated, we can now readily obtain equations for intercepts 3 and 4.

Since

$$\text{slope 1} = -\frac{\text{intercept 1}}{\text{intercept 3}} \tag{A2.23}$$

it follows that

$$\text{intercept 3} = -\frac{\text{intercept 1}}{\text{slope 1}} \tag{A2.24}$$

$$= -\frac{n_T \langle k \rangle_1}{-\langle k \rangle_2}$$

$$\text{intercept 3} = \frac{n_T \langle k \rangle_1}{\langle k \rangle_2} \tag{A2.25}$$

In turn to obtain intercept 4, we note that

$$\text{slope 2} = -\frac{\text{intercept 4}}{\text{intercept 2}} \tag{A2.26}$$

Hence

$$\text{intercept 4} = -\text{slope 2} \times \text{intercept 2} \tag{A2.27}$$

$$= -(-\langle k \rangle_0) n_T$$

$$\text{intercept 4} = n_T \langle k \rangle_0$$

$$= n_T \frac{n_1 + n_2}{\dfrac{n_1}{k_1} + \dfrac{n_2}{k_2}} \tag{A2.28}$$

Thus it is evident that for two classes of (identical) sites, the intercepts and limiting slopes are, in general, not the quantities one would assume by simplistic extrapolation from a single class. For example, in Figure A2.1, intercept 3 is *not* the number of sites in the first class of identical sites n_1 but is rather the parameter

$$\text{intercept 3} = \frac{(n_1 k_1 + n_2 k_2)^2}{n_1 k_1^2 + n_2 k_2^2} \tag{A2.29}$$

Only if $k_2 \ll k_1$, and n_2 is of the order of magnitude of n_1, will this function for intercept 3 approach n_1.

Similarly intercept 4 is *not* equal to k_2, the binding constant of the second class of identical sites, but rather the complicated function

$$\text{intercept 4} = \frac{(n_1 + n_2)^2 k_1 k_2}{n_1 k_2 + n_2 k_1} \tag{A2.30}$$

A concrete feeling for the significance of these graphical relations is perhaps best obtained by considering the special case of just two independent sites (Fig. A2.2):

$$n_1 = n_2 = 1 \tag{A2.31}$$

$$\text{intercept 1} = k_1 + k_2 = 2\langle k\rangle_1 \tag{A2.32}$$

$$\text{intercept 2} = 2 \tag{A2.33}$$

$$\text{intercept 3} = \frac{(k_1 + k_2)^2}{k_1^2 + k_2^2} \tag{A2.34}$$

$$\text{intercept 4} = \frac{4k_1k_2}{k_1 + k_2} \tag{A2.35}$$

$$\text{slope 1} = -\frac{k_1^2 + k_2^2}{k_1 + k_2} \tag{A2.36}$$

$$\text{slope 2} = -\frac{2k_1k_2}{k_1 + k_2} \tag{A2.37}$$

If, for example, we specify further that $k_2 = 0.5\ k_1$, then

$$\begin{aligned}
&\text{intercept 1} = 1.5k_1 \neq k_1\\
&\text{intercept 2} = 2\\
&\text{intercept 3} = 1.8\ (\neq 1)\\
&\text{intercept 4} = 2.66k_2\ (\neq k_2)
\end{aligned}$$

So far our discussion has focused on a receptor with two classes of binding sites. To generalize, we should turn to m classes of sites, each class with identical, invariant site-binding constants. The algebraic steps required are straightforward.

To obtain a generalized equivalent to equation A2.6 for intercept 1, we should extend our summation to m classes in place of just 2. Then

$$\text{intercept 1} = \sum_{j=1}^{m} n_jk_j = n_T \frac{\sum_1^m n_jk_j}{\sum_1^m n_j} = n_T\langle k\rangle_1 \tag{A2.38}$$

$$\text{intercept 2} = \sum_{j=1}^{m} \frac{n_jk_j}{k_j} = \sum_{1}^{m} n_j = n_T \tag{A2.39}$$

$$\text{slope } 1 = -\frac{\sum_{j=1}^{m} n_j k_j^2}{\sum_1^m n_j k_j} = -\langle k \rangle_2 \tag{A2.40}$$

$$\text{intercept } 3 = \frac{n_T \langle k \rangle_1}{\langle k \rangle_2} \tag{A2.41}$$

$$\text{slope } 2 = -\frac{\sum_{i=i}^{m} n_j}{\sum_{j=1}^{m} \frac{n_j}{k_j}} = -\langle k \rangle_0 \tag{A2.42}$$

$$\text{intercept } 4 = n_T \langle k \rangle_0 \tag{A2.43}$$

Finally, we turn to the alternative linearized graph of Figure A2.3 in which $1/B$ is plotted against $1/L$. Let us find the values for the intercepts and slopes for a receptor with m classes of sites, each class with identical, invariant binding constants.

We start with a generalized version of equation A2.1.

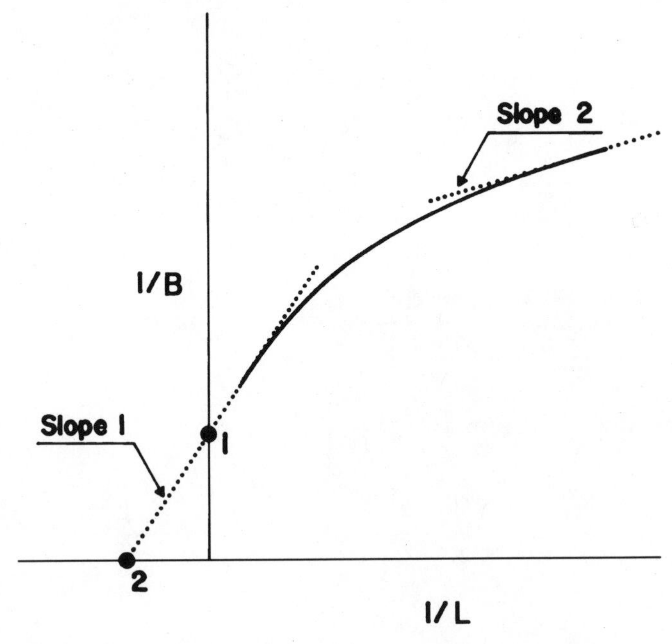

Figure A2.3 A schematic curve with intercepts and limiting slopes for a graph of variables $1/B$ and $1/L$ where m independent classes of sites are present.

$$B = \sum_{j=1}^{m} \frac{n_j k_j L}{1 + k_j L} \tag{A2.44}$$

From this it follows that

$$\frac{1}{B} = \left[\sum_{j=1}^{m} \frac{n_j k_j L}{1 + k_j L} \right]^{-1} = \left[\sum_{1}^{m} \frac{n_j k_j}{(1/L + k_j)} \right]^{-1} \tag{A2.45}$$

It is then possible by algebraic steps similar to those used to establish the limiting slopes and intercepts for Figure A2.1 to show that the following relations apply in Figure A2.3 (1):

$$\text{intercept 1} = \lim_{1/L \to 0} \left(\frac{1}{B} \right) = \left[\sum_{1}^{m} \frac{n_j k_j}{k_j} \right]^{-1} = \frac{1}{n_T} \tag{A2.46}$$

$$\text{slope 1} = \lim_{1/L \to 0} \left[\frac{d\left(\frac{1}{B}\right)}{d\left(\frac{1}{L}\right)} \right] = \frac{\sum_1^m \left(\frac{n_j}{k_j}\right)}{\left[\sum_1^m n_j\right]^2} = \frac{1}{n_T \langle k \rangle_0} \tag{A2.47}$$

$$\text{intercept 2} = -\frac{\text{intercept 1}}{\text{slope 1}} = -\langle k \rangle_0 \tag{A2.48}$$

$$\text{slope 2} = \lim_{L \to 0} \left[\frac{d(1/B)}{d(1/L)} \right]$$

$$= \frac{\sum_1^m n_j k_j}{\left[\sum_1^m n_j k_j\right]^2} = \frac{1}{\sum_1^m n_j k_j} = \frac{1}{n_T \langle k \rangle_1} \tag{A2.49}$$

Again, to provide a concrete feeling for the generalized graphical parameters in Figure A2.3, let us consider the special case of just two independent sites, $n_1 = n_2 = 1$. Under these circumstances, the following are obtained by appropriate substitutions in equations A2.46 to A2.49 (see Fig. A2.4):

$$\text{intercept } 1 = \frac{1}{n_T} = \frac{1}{2} \tag{A2.50}$$

$$\text{slope } 1 = \frac{k_1 + k_2}{4k_1k_2} \tag{A2.51}$$

$$\text{intercept } 2 = -\frac{2k_1k_2}{k_1 + k_2} \tag{A2.52}$$

$$\text{slope } 2 = \frac{1}{k_1 + k_2} \tag{A2.53}$$

If we become even more specific and assume that $k_2 = 0.5\ k_1$, then we find that

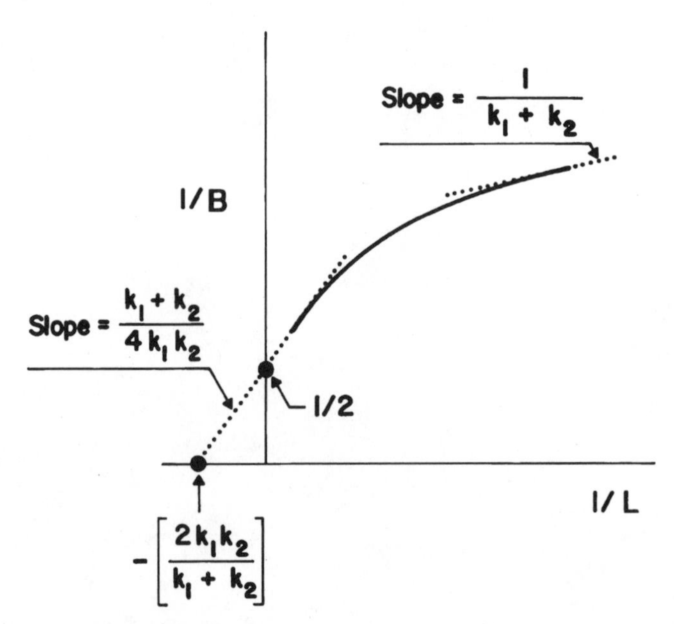

Figure A2.4 Schematic curve for a system of only two independent binding sites. The intercepts and limiting slopes are shown.

$$\text{intercept } 1 = \tfrac{1}{2}$$
$$\text{slope } 1 = 0.75k_1^{-1} \neq k_1^{-1}$$
$$\text{intercept } 2 = -0.67k_1 \neq k_1$$
$$\text{slope } 2 = 0.67k_1^{-1} \neq k_1^{-1}$$

Thus we see that neither the slopes nor the intercepts give a value for k_1 per se but rather lead to some fraction thereof.

REFERENCE

1. I. M. Klotz and D. L. Hunston, *Biochemistry* **10,** 3065–3069 (1971).

APPENDIX A3

RELATIONSHIPS BETWEEN STOICHIOMETRIC BINDING CONSTANTS AND SITE BINDING CONSTANTS

Stoichiometric and site binding constants must be interrelated since each group is connected to the observed dependence of moles bound ligand B on free ligand concentration L. When the binding sites are independent and invariant in affinity for ligand, then

$$\frac{K_1 L + 2K_1K_2L^2 + \cdots + n\left(\prod_1^n K_l\right) L^n}{1 + K_1 L + K_1K_2L^2 + \cdots + \left(\prod_1^n K_l\right) L^n}$$

$$= B = \frac{k_1 L}{1 + k_1 L} + \frac{k_2 L}{1 + k_2 L} + \cdots + \frac{k_n L}{1 + k_n L} \tag{A3.1}$$

Under these circumstances the number of stoichiometric constants K_i equals the number of site constants k_j. Since in equation A3.1 both sets of binding constants specify the same experimental quantity B, the individual K_i values must be uniquely connnected with k_j values. However, when the binding sites change their affinities with increasing occupancy by ligand, it is not possible to write an equation similar to the right-hand side of equation A3.1 containing all the occupancy-dependent site constants (Chapter 2). Nevertheless, one can find relations between K_i and certain combinations of occupancy-dependent site binding constants, even though these do not provide unique correlations.

Let us examine these situations in detail.

A3.1. DIVALENT SYSTEMS

A3.1.1. Independent Sites with Invariant Affinities

For a two-site receptor, the relationships between K_i and k_j were formulated 80 years ago (1,2). A simple procedure for finding the pertinent equations can be formulated by starting with a schematic diagram that visualizes the interrelationships between these constants (Fig. A3.1). The equilibrium pathways to the ligand–receptor series ${}_1RL$, ${}_2RL$, and ${}_{1,2}RL_2$, in which sites 1, 2, and both, respectively, are occupied by ligand, are shown with the corresponding site equilibrium constants. Below the site equilibria are schematic, abbreviated equations showing the species involved in the stoichiometric equilibria.

Since conservation of mass requires that

$$(RL_1) = ({}_1RL) + ({}_2RL) \tag{A3.2}$$

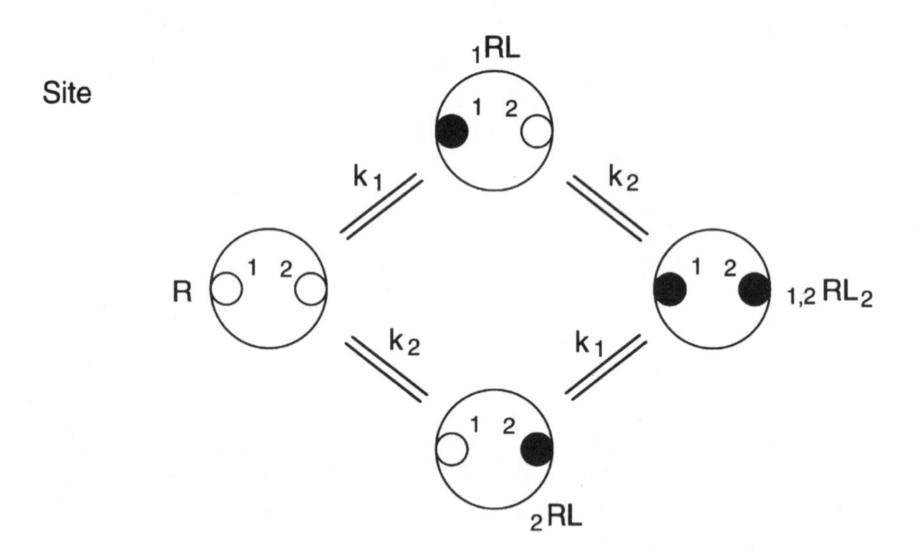

Figure A3.1 Comparison of meanings of stoichiometric and site binding constants for a two-site receptor in which the sites have different, invariant affinities.

we can write

$$K_1 = \frac{(RL_1)}{(R)(L)} = \frac{[({}_1RL) + ({}_2RL)]}{(R)(L)} = k_1 + k_2 \tag{A3.3}$$

In addition,

$$\begin{aligned} K_1K_2 &= \frac{(RL_1)}{(R)(L)} \frac{(RL_2)}{(RL_1)(L)} = \frac{(RL_2)}{(R)(L)^2} \\ &= \frac{(RL_2)}{(R)(L)^2} \frac{({}_1RL_1)}{({}_1RL_1)} = \frac{(RL_2)}{({}_1RL_1)(L)} \frac{({}_1RL_1)}{(R)(L)} \end{aligned} \tag{A3.4}$$

which leads to

$$K_1K_2 = k_1k_2 \tag{A3.5}$$

Thus if K_1 and K_2 have been established from experimental data for B as a function of L, equations A3.3 and A3.5 permit one to calculate corresponding values of k_1 and k_2 for a receptor, but only if it is known from *extrathermodynamic* information that the receptor has two sites with different invariant affinities.

A3.1.2. Site Affinities Change with Occupancy

If the site affinities change as ligand is increasingly bound, the situation becomes more complex. A schematic representation of the equilibria is presented in Figure A3.2. The site constants for the uptake of the second ligand now need a double index $k_{j_1j_2}$; the first j refers to the index number of the site first occupied and the second j gives the index number of the site occupied in the second step.

In Figure A3.2, we see that there are two alternative avenues for proceeding from ligand-free receptor R to fully saturated receptor ${}_{1,2}RL_2$. The first ligand could occupy site 1 to give ${}_1RL$; the corresponding site equilibrium constant is k_1. A second ligand can then occupy site 2 of ${}_1RL$ to give ${}_{1,2}RL_2$. For this second step, the site equilibrium constant is not k_2, for the latter is a measure of the affinity of site 2 in R, i.e., when site 1 is empty. However, in ${}_1RL$, site 1 is occupied and can exert an effect on the affinity for ligand of site 2. Therefore, the site equilibrium constant must be designated by an

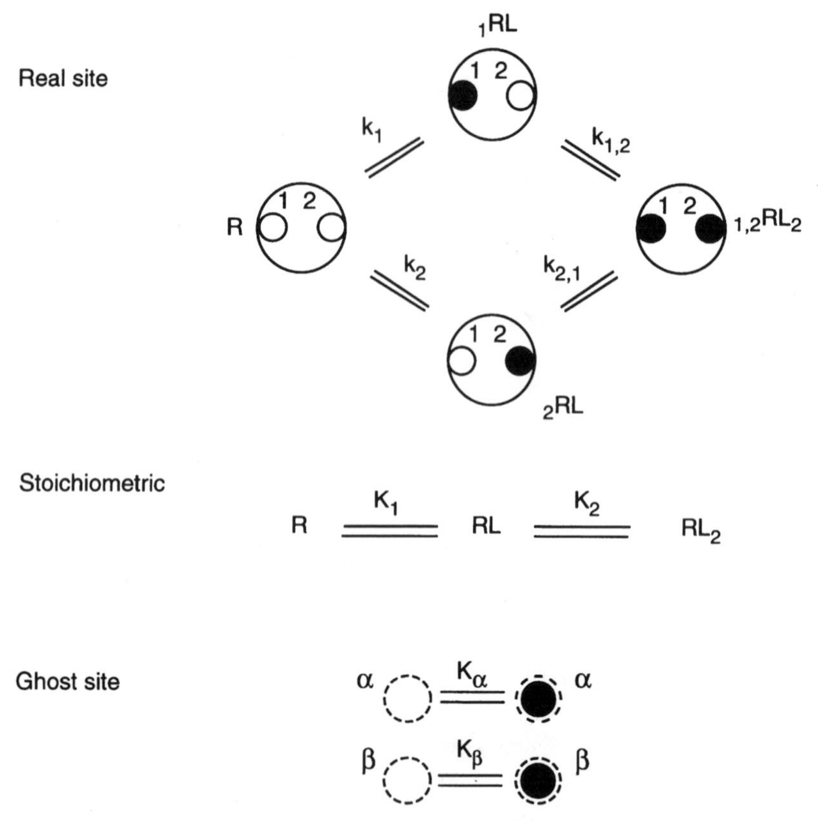

Figure A3.2 Comparison of meanings of different types of ligand–receptor binding constants for a two-site receptor in which site affinities may change with occupancy by ligand. Distinctions must be made between the stoichiometric formulation and the site formulations in terms of real or ghost sites.

index $k_{1,2}$ that clearly distinguishes it from k_2. For the alternative pathway from empty R to saturated $_{1,2}RL_2$ we define two site equilibrium constants k_2 and $k_{2,1}$.

Insofar as the stoichiometric equilibria are concerned, however, we need not take cognizance of the sites. For uptake of the first mole of ligand, we can define K_1, for the second mole, K_2 (see Fig. A3.2). From general algebraic arguments described earlier (Chapter 4), we also define two ghost-site equilibrium constants K_α and K_β for the corresponding ghost sites.

Thus it is evident that a bivalent ligand–receptor system has assigned to it four site constants but only two stoichiometric constants.

Actually, only three of the four site constants are independent. By inserting the appropriate molecular species into the product constants $k_1k_{1,2}$ and $k_2k_{2,1}$.

$$k_1k_{1,2} = \frac{({}_1RL)}{(R)(L)}\frac{({}_{1,2}RL_2)}{({}_1RL)(L)}; \quad k_2k_{2,1} = \frac{({}_2RL)}{(R)(L)}\frac{({}_{1,2}RL_2)}{({}_2RL)(L)} \tag{A3.6}$$

we can see that

$$k_1k_{1,2} = k_2k_{2,1} \tag{A3.7}$$

Thus if any three of the site constants are known, the fourth can be calculated. Nevertheless, three out of the four are independent parameters.

We can still find relations between the stoichiometric equilibrium constants and the site ones. Thus for the first stoichiometric step

$$K_1 = \frac{(RL)}{(R)(L)} = \frac{[({}_1RL) + ({}_2RL)]}{(R)(L)} = k_1 + k_2, \tag{A3.8}$$

and for the second

$$K_1K_2 = \frac{(RL_2)}{(R)(L)^2} = \frac{(RL_2)}{(R)(L)^2}\frac{({}_1RL)}{({}_1RL)} = \frac{({}_1RL)}{(R)(L)}\frac{(RL_2)}{({}_1RL)(L)} = k_1k_{1,2} \tag{A3.9}$$

However, now we are faced with an indeterminacy in the site constants. The stoichiometric constants K_1 and K_2 could be evaluated for a given set of binding data by fitting them to the stoichiometric binding equation A3.1. However, there is no way of specifying all three site constants from the two equations A3.8 and A3.9. A third relation between site constants is supplied by equation A3.7, but simultaneously it introduces a fourth variable. Thus from binding data alone there is no way of establishing values for the three independent site constants; they can exhibit free will, i.e., one of them can assume any value as long as the other two are constrained by the relations of equations A3.8 and A3.9.

For a divalent receptor for which site affinities change with occupancy by ligand, the following equation (Chapter 4)

$$\frac{K_1L + 2K_1K_2L^2}{1 + K_1L + K_1K_2L^2} = B = \frac{K_\alpha L}{1 + K_\alpha L} + \frac{K_\beta L}{1 + K_\beta L} \tag{A3.10}$$

is fully valid, with ghost-site binding constants K_α and K_β. However, K_α or K_β is *not* simply k_1 or k_2 or $k_{1,2}$ or $k_{2,1}$.

A3.1.3. Ideal Situation: All Sites Identical with Invariant Affinities Under ideal circumstances, equations A3.3 and A3.5 can be reduced to

$$K_1 = k + k \tag{A3.11}$$

$$K_1K_2 = kk \tag{A3.12}$$

Thus

$$K_1 = 2k \tag{A3.13}$$

$$K_2 = \frac{k}{2} \tag{A3.14}$$

in agreement with equations 6.24 and 6.26, which were derived by a different procedure.

A3.2. Trivalent Systems

If we turn to a trivalent ligand–receptor complex, the number of sites increases to 12 (Fig. A3.3). If the sites have different but invariant affinities, then

$$k_1 = k_{2,1} = k_{2,3,1} \tag{A3.15}$$

$$k_2 = k_{1,2} = k_{3,1,2} \tag{A3.16}$$

$$k_3 = k_{2,3} = k_{1,2,3} \tag{A3.17}$$

and the system is defined by only three independent site constants. Their relationships to the stoichiometric constants can be derived by a procedure similar to that used for divalent systems.

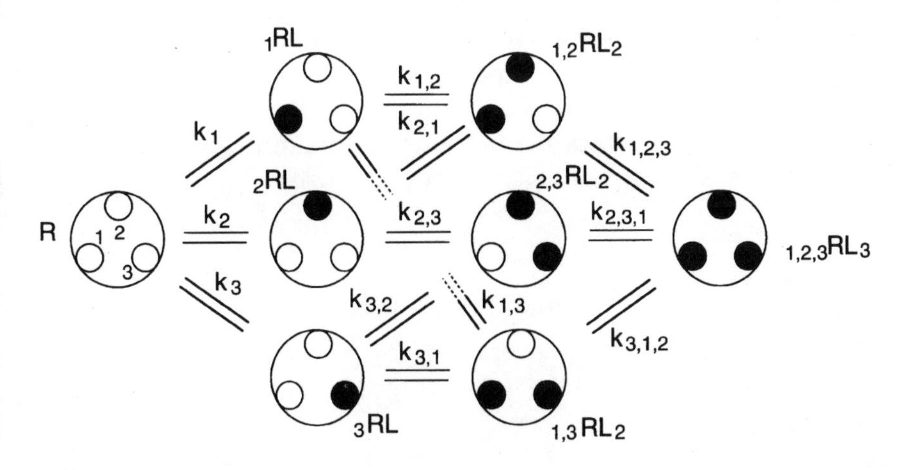

$$R \overset{K_1}{=\!=\!=} RL_1 \overset{K_2}{=\!=\!=} RL_2 \overset{K_3}{=\!=\!=} RL_3$$

Figure A3.3 Comparison of meanings of different types of ligand–receptor binding constants for a three-site receptor in which the site affinities change with extent of occupancy of receptor by ligand molecules.

For K_1 we write

$$K_1 = \frac{(RL_1)}{(R)(L)} = \frac{[({}_1RL) + ({}_2RL) + ({}_3RL)]}{(R)(L)} = k_1 + k_2 + k_3 \qquad \text{(A3.18)}$$

Correspondingly, to go from R to RL_2 we must put on two ligands in succession to form site-occupied species ${}_{1,2}RL_2$, also two to form ${}_{1,3}RL_2$, and then two to form ${}_{2,3}RL_2$. Each of these two-site filling steps is governed by a product site equilibrium constant, k_1k_2, k_1k_3 and k_2k_3, respectively. Consequently, we can show that

$$K_1K_2 = \frac{(RL_2)}{(R)(L)^2} = \frac{[({}_{1,2}RL_2) + ({}_{1,3}RL_2) + ({}_{2,3}RL_2)]}{(R)(L)^2}$$
$$= k_1k_2 + k_1k_3 + k_2k_3 \qquad \text{(A3.19)}$$

Finally, to go from R to RL_3 we must put on three ligands in succession. The order of filling the three sites does not matter since the final result in each path is the same, ${}_{1,2,3}RL_3$. (The free energy change ΔG° for a transformation depends only on the initial and final states not on the path used to go between them.) Thus we can take the ordinal sequential path, in which the overall equilibrium is governed by the site constant $k_1k_2k_3$. Consequently, we can write

$$\begin{aligned} K_1K_2K_3 &= \frac{(RL_3)}{(R)(L)^3} = \frac{[({}_{1,2,3}RL_3)]}{(R)(L)^3} \\ &= \frac{({}_{1,2,3}RL_3)}{(R)(L^3)} \; \frac{({}_{1,2}RL_2)}{({}_{1,2}RL_2)} \; \frac{({}_1RL_1)}{({}_1RL_1)} \\ &= \frac{({}_1RL_1)}{(R)(L)} \; \frac{({}_{1,2}RL_2)}{({}_1RL_1)(L)} \; \frac{({}_{1,2,3}RL_3)}{({}_{1,2}RL_2)(L)} \end{aligned} \tag{A3.20}$$

Under ideal circumstances all of the site constants are identical and invariant in affinity and can be represented by a single k. Equations A3.18 and A3.20 can then be reduced to the following (statistical) relationships:

$$K_1 = 3k \tag{A3.21}$$

$$K_2 = k \tag{A3.22}$$

$$K_3 = \frac{k}{3} \tag{A3.23}$$

A3.3. TETRAVALENT SYSTEM

The appropriate stoichiometric and site binding constants for a four-site complex, such as occurs with hemoglobin, are defined in Figure A3.4. In the most general circumstances, 32 different site binding constants exist. If the sites have different but invariant affinities, then the site constant for filling site 1 is always k_1 and those for filling sites 2, 3, and 4 are always k_2, k_3, and k_4, respectively. So there are only four independent constants. In this situation, their relationships to stoichiometric constants can be derived by a procedure similar to that shown for divalent and trivalent systems. Without going through the algebraic steps, one can see that these interrelationships should be expressed as follows:

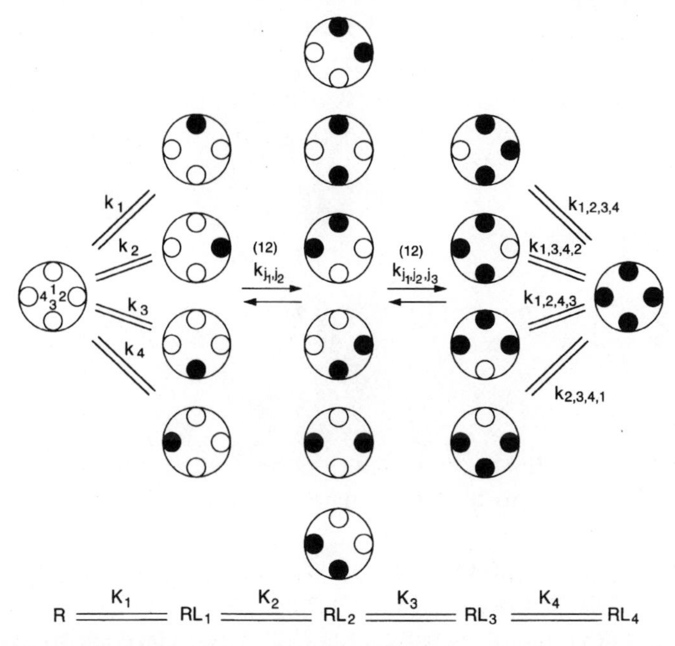

Figure A3.4 Stoichiometric and site binding constants for a four-site receptor in which site affinities change with extent of occupancy of receptor by ligand molecules.

$$K_1 = k_1 + k_2 + k_3 + k_4 = \sum_{1}^{4} k_{j_i} \tag{A3.24}$$

$$K_1K_2 = k_1k_2 + k_1k_3 + k_1k_4 + k_2k_3 + k_2k_4 + k_3k_4 \tag{A3.25}$$

$$= \sum_{j_1=1}^{4-1} \sum_{j_2=j_1+1}^{4} k_{j_1}k_{j_2}$$

$$K_1K_2K_3 = k_1k_{1,2}k_{1,2,3} + k_1k_{1,2}k_{1,2,4} + k_1k_{1,3}k_{1,3,4} + k_2k_{2,3}k_{2,3,4} \tag{A3.26}$$

$$= \sum_{j_1=1}^{4-2} \sum_{j_2=j_1+1}^{4-1} \sum_{j_3=j_2+1}^{4} k_{j_1}k_{j_2}k_{j_3}$$

$$K_1K_2K_3K_4 = k_1k_{1,2}k_{1,2,3}k_{1,2,3,4} \tag{A3.27}$$

Under ideal circumstances, all of the site constants are identical and invariant in affinity so they can be represented by a single k. Equations A3.24 to A3.27 can then be reduced to the following (statistical) relationships:

$$K_1 = 4k \tag{A3.28}$$

$$K_2 = \tfrac{3}{2}\,k \tag{A3.29}$$

$$K_3 = \tfrac{2}{3}\,k \tag{A3.30}$$

$$K_4 = \tfrac{1}{4}\,k \tag{A3.31}$$

If all the sites in hemoglobin had identical unchanging affinities for O_2, the stoichiometric binding constants would have the relative values shown.

If the site constants change with increasing uptake of ligand, the equations relating them to the stoichiometric constants K_i are made up of many independent parameters and become too unwieldy to provide general insights.

A3.4. MULTISITE SYSTEM

The relationships between K_i values and k_j values have been deduced by different procedures (3). Their general format can be seen by stepwise extrapolation of the equations shown above for divalent, trivalent, and tetravalent complexes, and can be written as follows:

$$K_1 = k_1 + k_2 + \cdots + k_n = \sum_{j_i=1}^{n} k_{j_1} \tag{A3.32}$$

$$K_1K_2 = k_1k_2 + k_1k_3 + \cdots + k_1k_n + k_2k_3 + k_2k_4 + \cdots + k_2k_n + \cdots$$

$$= \sum_{j=1}^{n-1} \sum_{j_2=j_1+1}^{n} k_{j_1}k_{j_2} \tag{A3.33}$$

$$\vdots$$

$$K_1K_2 \cdots K_i = \sum_{j_1=1}^{n-i+1} \sum_{j_2=j_1+1}^{n-i+2} \cdots \sum_{j_i=j_{i-1}+1}^{n} k_{j_1}k_{j_2} \cdots k_{j_i}$$

$$\vdots \tag{A3.34}$$

$$K_1K_2 \cdots K_n = k_1k_2 \cdots k_n \tag{A3.35}$$

Under ideal circumstances all the site constants are identical and invariant in affinity so they can be represented by a single k. Equations A3.32 to A3.35 can then be reduced to the following (statistical) relationships:

$$K_1 = nk \tag{A3.36}$$

$$K_2 = \frac{n-1}{2} k \tag{A3.37}$$

$$\vdots$$

$$K_i = \frac{n-i+1}{i} k \tag{A3.38}$$

$$\vdots$$

$$K_n = \frac{1}{n} k \tag{A3.39}$$

These equations, and the normalized form

$$iK_i = (n+1)k - ki \tag{A3.40}$$

provide a format for comparing actual binding behavior with the ideal for a multisite ligand–receptor complex.

REFERENCES

1. E. Q. Adams, *J. Am. Chem. Soc.* **38,** 1503–1510 (1916).
2. H. S. Simms, *J. Am. Chem. Soc.,* **48,** 1230–1261 (1926).
3. I. M. Klotz and D. L. Hunston, *J. Biol. Chem.* **250,** 3001–3009 (1975), and references therein.

APPENDIX A4

RELATIONSHIPS BETWEEN STOICHIOMETRIC BINDING CONSTANTS AND GHOST-SITE BINDING CONSTANTS

To obtain the relations between stoichiometric and ghost-site binding constants, we start with the polynomial, equation 4.1, which represents the partition function for a receptor binding a ligand,

$$\begin{aligned} Z &= 1 + K_1L + K_1K_2L^2 + \cdots + (K_1 \cdots K_n)L^n \\ &= (K_1 \cdots K_n)[(L - a_1)(L - a_2) \cdots (L - a_n)] \end{aligned} \tag{A4.1}$$

In this equation, K_1, K_2, etc. are the stoichiometric equilibrium constants, L is the concentration of free nonbound ligand, and a_1, a_2, etc. are the roots of the polynomial.

A4.1. DIVALENT SYSTEMS

When 2 mol of bound ligand saturate the receptor, n in equation A4.1 is 2, and there are only two roots of the polynomial. Hence we may write

$$Z = 1 + K_1L + K_1K_2L^2 = K_1K_2(L - a_1)(L - a_2) \tag{A4.2}$$

If we expand the two factors for the roots we find

$$(L - a_1)(L - a_2) = L^2 + [(-a_1) + (-a_2)]L + (-a_1)(-a_2) \qquad \text{(A4.3)}$$

Consequently from equation A4.1 we conclude that

$$\frac{Z}{K_1K_2} = Z' = \frac{1}{K_1K_2} + \frac{K_1}{K_1K_2} L + \frac{K_1K_2}{K_1K_2} L^2$$

$$= (-a_1)(-a_2) + [(-a_1) + (-a_2)]L + L^2 \qquad \text{(A4.4)}$$

As pointed out in Chapter 4, we can replace each root a_λ by an alternative constant

$$K_\lambda = -\frac{1}{a_\lambda} \qquad \text{(A4.5)}$$

For a divalent system,*

$$K_\alpha = -\frac{1}{a_1} \qquad \text{(A4.6)}$$

$$K_\beta = -\frac{1}{a_2} \qquad \text{(A4.7)}$$

Then equation A4.4 can be transformed into

$$\frac{1}{K_1K_2} + \frac{K_1}{K_1K_2} L + L^2 = \frac{1}{K_\alpha K_\beta} + \frac{K_\alpha + K_\beta}{K_\alpha K_\beta} L + L^2 \qquad \text{(A4.8)}$$

A necessary and sufficient condition that these two different binomials be the same (each equal to Z') is that corresponding coefficients of

*The difference in indexing in equations A4.6 and A4.7 may puzzle some readers. It has long been customary to designate the roots of a polynomial with ordinal Arabic subscripts, i.e., a_1, a_2, etc. The corresponding ghost-site equilibrium constants cannot be assigned parallel subscript numbers because K_1K_2, etc. have been preempted as symbols for stoichiometric equilibrium constants. Virtual or ghost-site equilibrium constants have been designated, therefore, by subscripts from the Greek alphabet, α corresponding to 1; β, to 2; etc., with K being retained because these constants are linked algebraically to the stoichiometric ones (see equation 4.1). Similar considerations apply when ghost-site equilibrium constants appear in summations.

each L^i term be equal. Consequently we can conclude that

$$K_1 K_2 = K_\alpha K_\beta \tag{A4.9}$$

$$K_1 = K_\alpha + K_\beta \tag{A4.10}$$

$$K_2 = \frac{K_\alpha K_\beta}{K_\alpha + K_\beta} \tag{A4.11}$$

Thus once the stoichiometric binding constants are known, one can compute values for the virtual or ghost-site binding constants.

For a divalent complex one can also readily obtain explicit analytic expressions for K_α and K_β in terms of K_1 and K_2. Using equation A4.9 to replace K_β in equation A4.10 and making a simple rearrangement, one forms a quadratic equation in K_α,

$$(K_\alpha)^2 - K_1(K_\alpha) + K_1 K_2 = 0 \tag{A4.12}$$

The solution to this equation is

$$K_\alpha = \tfrac{1}{2}K_1 \pm \tfrac{1}{2}(K_1^2 - 4K_1K_2)^{1/2} \tag{A4.13}$$

From this and equation A4.10 it follows that

$$K_\beta = \tfrac{1}{2}K_1 \mp \tfrac{1}{2}(K_1^2 - 4K_1K_2)^{1/2} \tag{A4.14}$$

It should be noted immediately that when $K_2 > \frac{1}{4}K_1$, which would be true if affinities increased with extent of occupancy (see equations 6.24, 6.26), K_α and K_β will have complex values.

For a divalent complex it is also possible to derive explicit expressions for K_α and K_β in terms of site binding constants:

$$K_\alpha = \tfrac{1}{2}(k_1 + k_2) \pm \tfrac{1}{2}[(k_1 + k_2)^2 - 4k_1k_{1,2}]^{1/2} \tag{A4.15}$$

$$K_\beta = \tfrac{1}{2}(k_1 + k_2) \mp \tfrac{1}{2}[(k_1 + k_2)^2 - 4k_1k_{1,2}]^{1/2} \tag{A4.16}$$

Note that K_α is *not* identifiable with any specific site constant, nor is K_β.

A4.2. TRIVALENT SYSTEMS

When 3 mol of bound ligand saturate the receptor, n in equation A4.1 is 3 and there are three roots of the polynomial Z. If we expand the three factors for the roots, we find

$$\begin{aligned}(L - a_1)(L - a_2)(L - a_3) \\ = L^3 + [(-a_1) + (-a_2) + (-a_3)]L^2 + [(-a_1)(-a_2) + (-a_1)(-a_3) \\ + (-a_2)(-a_3)]L + [(-a_1)(-a_2)(-a_3)]\end{aligned} \tag{A4.17}$$

For a trivalent complex, equation A4.1 becomes

$$\begin{aligned}Z &= 1 + K_1L + K_1K_2L^2 + K_1K_2K_3L^3 \\ &= K_1K_2K_3[\{(-a_1)(-a_2)(-a_3)\} \\ &\quad + \{(-a_1)(-a_2) + (-a_1)(-a_3) + (-a_2)(-a_3)\}L \\ &\quad + \{(-a_1) + (-a_2) + (-a_3)\}L^2 + L^3]\end{aligned} \tag{A4.18}$$

If we replace the roots by the corresponding ghost-site constants

$$K_\alpha = -\frac{1}{a_1} \tag{A4.19}$$

$$K_\beta = -\frac{1}{a_2} \tag{A4.20}$$

$$K_\gamma = -\frac{1}{a_3} \tag{A4.21}$$

we can transform equation A4.18 into

$$\begin{aligned}&\frac{1}{K_1K_2K_3} + \frac{K_1}{K_1K_2K_3}L + \frac{K_1K_2}{K_1K_2K_3}L^2 + \frac{K_1K_2K_3}{K_1K_2K_3}L^3 \\ &= \frac{1}{K_\alpha K_\beta K_\gamma} + \left\{\frac{1}{K_\alpha K_\beta} + \frac{1}{K_\alpha K_\gamma} + \frac{1}{K_\beta K_\gamma}\right\}L \\ &\quad + \left\{\frac{1}{K_\alpha} + \frac{1}{K_\beta} + \frac{1}{K_\gamma}\right\}L^2 + L^3\end{aligned} \tag{A4.22}$$

If we equate coefficients of each of the same L^i terms on both sides of equation A4.22, we find that

$$K_1 K_2 K_3 = K_\alpha K_\beta K_\gamma \tag{A4.23}$$

$$\frac{K_1}{K_1 K_2 K_3} = \frac{1}{K_\alpha K_\beta} + \frac{1}{K_\alpha K_\gamma} + \frac{1}{K_\beta K_\gamma} = \frac{K_\alpha + K_\beta + K_\gamma}{K_\alpha K_\beta K_\gamma} \tag{A4.24}$$

$$\frac{K_1 K_2}{K_1 K_2 K_3} = \frac{1}{K_\alpha} + \frac{1}{K_\beta} + \frac{1}{K_\gamma} = \frac{K_\alpha K_\beta + K_\alpha K_\gamma + K_\beta K_\gamma}{K_\alpha K_\beta K_\gamma} \tag{A4.25}$$

Placed in a more ordered array, these equations became

$$K_1 = K_\alpha + K_\beta + K_\gamma = \sum_{1}^{3} K_{\lambda_1} \tag{A4.26}$$

$$K_1 K_2 = K_\alpha K_\beta + K_\alpha K_\gamma + K_\beta K_\gamma = \sum_{\lambda_1 = 1}^{3-1} \sum_{\lambda_2 = \lambda_1 + 1}^{3} K_{\lambda_1} K_{\lambda_2} \tag{A4.27}$$

$$K_1 K_2 K_3 = K_\alpha K_\beta K_\gamma \tag{A4.28}$$

Thus, again, once the stoichiometric binding constants are known, one can compute values for the ghost-site binding constants. However, in practice it is arduous to find explicit analytical solutions to cubic equations, which are unwieldy, and hence it is much more convenient to use numerical methods (1).

A4.3. TETRAVALENT SYSTEMS

When 4 mol of bound ligand saturate the receptor, n in equation A4.1 is 4, and there are four roots of the polynomial Z. If we expand the four factors for the roots, we find

$$
\begin{aligned}
(L - a_1)(L - a_2)(L - a_3)(L - a_4) \\
&= L^4 + [(-a_1) + (-a_2) + (-a_3) + (-a_4)]L^3 \\
&\quad + [(-a_1)(-a_2) + (-a_1)(-a_3) + (-a_1)(-a_4) + (-a_2)(-a_3) \\
&\quad + (-a_2)(-a_4) + (-a_3)(-a_4)]L^2 \\
&\quad + [(-a_1)(-a_2)(-a_3) + (-a_1(-a_2)(-a_4) + (-a_1)(-a_3)(-a_4) \\
&\quad + (-a_2)(-a_3)(-a_4)]L \\
&\quad + [(-a_1)(-a_2)(-a_3)(-a_4)] \qquad \text{(A4.29)}
\end{aligned}
$$

Carrying out algebraic manipulations analogous to those illustrated for divalent and trivalent systems, we find that

$$K_1 + K_\alpha + K_\beta + K_\gamma + K_\delta = \sum_{\lambda_1 = 1}^{4} K_{\lambda_1} \qquad \text{(A4.30)}$$

$$
\begin{aligned}
K_1 K_2 &= K_\alpha K_\beta + K_\alpha K_\gamma + K_\alpha K_\delta + K_\beta K_\gamma + K_\beta K_\delta + K_\gamma K_\delta \\
&= \sum_{\lambda_1 = 1}^{4-1} \sum_{\lambda_2 = \lambda_1 + 1}^{4} K_{\lambda_1} K_{\lambda_2} \qquad \text{(A4.31)}
\end{aligned}
$$

$$
\begin{aligned}
K_1 K_2 K_3 &= K_\alpha K_\beta K_\gamma + K_\alpha K_\beta K_\delta + K_\alpha K_\gamma K_\delta + K_\beta K_\gamma K_\delta \\
&= \sum_{\lambda_1 = 1}^{4-2} \sum_{\lambda_2 = \lambda_1 + 1}^{4-1} \sum_{\lambda_3 = \lambda_2 + 1}^{4} K_{\lambda_1} K_{\lambda_2} K_{\lambda_3} \qquad \text{(A4.32)}
\end{aligned}
$$

$$K_1 K_2 K_3 K_4 = K_\alpha K_\beta K_\gamma K_\delta \qquad \text{(A4.33)}$$

In practice, for tetravalent systems also, procedures for solving the quartic equations necessary to obtain explicit equations for K_{λ_i} in terms of K_i values are unwieldy, so numerical analysis devices are used (1).

A4.4. MULTIVALENT SYSTEMS

It is probably evident from the preceding specific examples that for an n-valent system, for which

$$1 + K_1L + K_1K_2L^2 + \cdots + (K_1 \cdots K_i)L^i + \cdots + (K_1 \cdots K_n)L^n$$
$$= (K_1 \cdots K_n)[(L - a_1)(L - a_2) \cdots (L - a_n)] \tag{A4.34}$$

algebraic expansions and transformations should lead to the following general expressions:

$$K_1 = \sum_{\lambda_1 = 1}^{n} K_{\lambda_1} \tag{A4.35}$$

$$K_1K_2 = \sum_{\lambda_1 = 1}^{n-1} \sum_{\lambda_2 = \lambda_1 + 1}^{n} K_{\lambda_1}K_{\lambda_2} \tag{A4.36}$$

$$\vdots$$

$$K_1K_2 \cdots K_i = \sum_{\lambda_1 = 1}^{n-i+1} \sum_{\lambda_2 = \lambda_1 + 1}^{n-i+2} \cdots \sum_{\lambda_i = \lambda_{i-1} + 1}^{n} K_{\lambda_1}K_{\lambda_2} \cdots K_{\lambda_i} \tag{A4.37}$$

$$\vdots$$

$$K_1K_2 \cdots K_n = K_\alpha K_\beta \cdots K_\nu \; (\nu = n) \tag{A4.38}$$

For polynomials of a degree higher than 4, it is not possible to find an analytic solution for K_{λ_i} in terms of K_i values. For all ligand–receptor complexes higher than quadrivalent, virtual or ghost-site constants are calculated from stoichiometric constants by numerical methods.

REFERENCES

1. I. M. Klotz, *Proc. Natl. Acad. Sci. USA*, **90,** 7191–7194 (1993).

INDEX